Technological Advance

in

An Expanding Economy:

Its Impact on a Cross-Section of the Labor Force

EVA MUELLER

with

JUDITH HYBELS
JAY SCHMIEDESKAMP
JOHN SONQUIST
CHARLES STAELIN

SURVEY RESEARCH CENTER

ISR

INSTITUTE FOR SOCIAL RESEARCH
THE UNIVERSITY OF MICHIGAN
ANN ARBOR, MICHIGAN

This report was prepared for the Manpower Administration, U.S. Department of Labor, under research contract No. 81-24-67-02 authorized by Title I of the Manpower Development and Training Act. Since contractors performing research under Government sponsorship are encouraged to express their own judgment freely, the report does not necessarily represent the Department's official opinion or policy. Moreover, the contractor is solely responsible for the factual accuracy of all material developed in the report.

Printed by Braun-Brumfield, Inc., Ann Arbor, Michigan
Manufactured in the United States of America

PREFACE

This monograph is a report on a nationwide survey, conducted by the Survey Research Center of The University of Michigan in 1967. It deals with the impact of changes in machine technology on a cross-section of the U.S. labor force. Three aspects of technological change are of particular concern in this study: (1) the economic impact of machine change on the work force in terms of income change, promotions, steadiness of employment and unemployment; (2) the relevance of machine change for job satisfaction and job content; and (3) the relation of machine change to education and training. Thus the subject matter of this study is not new. What *is* new is the scope of the study and the methodology employed. Previous investigations have been based on case studies and occasionally on indirect inferences from movements in broad aggregates. The results have often been of uncertain meaning, and even contradictory. This is not surprising, for the impact of change in machine technology is widely diffused and may differ greatly from case to case. A survey of the entire labor force can lead to generalizations and statements about the relative frequency of various consequences which case studies do not permit the researcher to make. However, the generality which commends a cross-sectional approach is achieved at some sacrifice of precision and detail. Hence case studies and cross-sectional studies must complement each other in the quest for an understanding of the impact of technological change.

The lack so far of cross-sectional studies of technological change is readily explained by the difficulty of studying so heterogeneous a phenomenon as change in machine technology by means of personal interviews. It was clear from the outset that this study would be almost like an experiment. The Survey Research Center is indebted to the Office of Manpower Research, Manpower Administration, U.S. Department of Labor for encouraging it to explore the cross-sectional approach. The Office of Manpower Research provided the financing for this project as well as valuable advice and suggestions.

iii

We gratefully acknowledge the substantive assistance extended by Dr. Howard Rosen, Director of Research, and members of his staff. In particular the study benefited from close collaboration with Sheridan Maitland of the Office of Manpower Research.

The author also wishes to acknowledge the indispensable assistance of her colleagues at the Survey Research Center. The Center is a division of the Institute for Social Research, which is directed by Rensis Likert. The Director of the Survey Research Center is Angus Campbell. The sample design and analysis of sampling error were the responsibility of Irene Hess. The field work was directed by Charles Cannell and John Scott, and the coding by Joan Scheffler. John Sonquist participated in the planning of the research and in the design of the questionnaire. Jay Schmiedeskamp saw the study through the data collection and coding stage. Judith Hybels and Charles Staelin participated in the data analysis and the writing of this monograph. William Haney prepared the monograph for publication. The author has worked for many years with George Katona, James Morgan and John Lansing. The intellectual influence of these three colleagues in the Economic Behavior Program of the Survey Research Center pervades this study.

Sue Hudson was responsible for typing of the tables, and Virginia Eaton undertook the typing of the draft manuscript.

The research reported in this monograph represents a large and daring departure from currently available studies of "automation." The cross-sectional approach cannot fill all the conspicuous gaps in our knowledge about labor force adjustment to technological change. Nor will all the estimates that it does yield be absolutely precise. A pioneering venture seldom attains immediate perfection. This holds for the present study. It is to be hoped that this survey will challenge further efforts along similar lines, and that improvements will come with experience.

CONTENTS

PREFACE   *iii*

TABLES   *viii*

CHAPTER 1 INTRODUCTION   *1*
  A. Defining Change in Machine Technology .3
  B. The Reference Period 5
  C. Advantages of the Cross-Sectional Approach 6
  D. Problems and Limitations of the Cross-Sectional Approach. 7
  E. Some Major Conclusions 10

CHAPTER 2 MACHINERY USE BY THE WORK FORCE   *17*
  A. The Frequency of Machinery Use 18
  B. Machinery Use in Relation to Socio-Economic Characteristics
     of the Labor Force 24
  C. People's Attitudes Toward Machinery .31

CHAPTER 3 CHANGE IN MACHINE TECHNOLOGY   *39*
  A. The Measurement of Change in Machine Technology 39
  B. Frequency of Experiences with Change in Machine Technology 42
  C. Kinds of Technological Change 47
  D. Who Experienced Changes in Machine Technology? 51

CHAPTER 4 THE PROCESS OF CHANGE-OVER
            TO THE NEW EQUIPMENT   *59*
  A. Advance Planning 60
  B. Training 63
  C. Transitional Unemployment 65
  D. Pay Changes 70
  E. Conclusion 72

CHAPTER 5   *ECONOMIC CONSEQUENCES OF*
            CHANGE IN MACHINE TECHNOLOGY   *75*
   A. Machine Use and Economic Situation of the Worker 76
   B. Machine Change and Change in Workers' Economic Situation 83
   C. Multivariate Analysis 93
   D. Job and Geographic Mobility in Relation to Machine Change 104

CHAPTER 6   *THE IMPACT OF CHANGES IN MACHINE*
            TECHNOLOGY ON PERCEIVED JOB CHAR-
            ACTERISTICS AND JOB SATISFACTION   *107*
   A. The Data Used 108
   B. Change in Job Satisfaction and Some of Its Determinants 111
   C. How Machine Change Alters Perceived Job Content and
      Job Characteristics 119
   D. Group Differences in Attitudes Toward Job Challenge 125

CHAPTER 7   *FURTHER ANALYSIS OF CASES*
            WHO MADE A POOR ADJUSTMENT TO
            TECHNOLOGICAL CHANGE   *135*
   A. The Adjustment Scale 136
   B. Statistical Analysis of Workers Who Made a Poor Adjustment 138
   C. Case Studies of Workers Who Made a Poor Adjustment 147

CHAPTER 8   *THE ROLE OF EDUCATION IN RELATION*
            TO TECHNOLOGICAL CHANGE   *159*
   A. Level of Machine Technology and Education 161
   B. Education and Attitudes Toward Machine Technology 166
   C. Felt Need for Education 170
   D. Education and Adjustment to Change 175
   E. Conclusions 178

CHAPTER 9   *FURTHER STEPS*   *181*

APPENDIX I   *INTERVIEWING AND QUESTIONNAIRE*
             DESIGN   *185*

APPENDIX II *SAMPLING METHODS AND*
             SAMPLING VARIABILITY   *231*
   A. The Sample 231
   B. Sampling Variability 232
   C. Comparisons With Independent Estimates 237

APPENDIX III DEFINITION AND MEASUREMENT
                OF AUTOMATION LEVEL   *241*
      A. The Conceptual Scheme 241
      B. Implementation 245

APPENDIX IV THE MULTIPLE CLASSIFICATION
                ANALYSIS   *247*
BIBLIOGRAPHY   *251*

LIST OF TABLES

CHAPTER 2

Table 2-1: Proportion of Members of the Labor Force Who Operate Machinery or Equipment on Their Jobs, by Sex . . 19

Table 2-2: Proportion of People Whose Work is Significantly Affected By Machinery or Equipment, by Sex 21

Table 2-3: Type and Automation Level of Machinery Important to Workers' Jobs 23

Table 2-4: Demographic Characteristics of People with Different Patterns of Equipment Use 25

Table 2-5: Automation Level of Equipment Used in Different Occupations 28

Table 2-6: Automation Level of Equipment Associated with Some Demographic Characteristics 30

Table 2-7: Attitudes Toward Automation by Pattern of Equipment Use and by Automation Level of Equipment 33

Table 2-8: Attitude Toward Job by Pattern of Equipment Use and by Automation Level of Equipment 34

Table 2-9: Workers' Perceptions of Their Opportunity to Talk and to Organize Their Work, by Pattern of Equipment Use 36

Table 2-10: Perceived Use of Skills by Workers, in Relation to Equipment Use 37

CHAPTER 3

Table 3-1: Distribution of the Labor Force by Changes in Machinery and Jobs over the Past Five Years 43

Table 3-2: Comparison of All Machinery Used and Machinery That Was Changed 48

Table 3-3: Labor Requirements of Changed Equipment 50

Table 3-4: Comparison of Machinery Used before and after Job Changes and Transfers 50

Table 3-5:   Automation Level of Equipment Operated by People Who
             Changed Jobs 52

Table 3-6:   Distribution of Workers by Age and Job and Machine
             Change During the Past 5 Years 52

Table 3-7:   Distribution of Workers by Education and Job and
             Machine Change During the Past 5 Years 53

Table 3-8:   Distribution of Workers by Occupation, Industry, and
             Job and Machine Change 55

Table 3-9:   Distribution of Workers by Size of Firm and Change in
             Firm's Employment and Job and Machine Change 56

CHAPTER 4

Table 4-1:   Planning the Change-Over: Advance Notice of Machine
             Change 61

Table 4-2:   Training at the Time of the Change-Over to New
             Machinery 64

Table 4-3:   Factors Which Were of the Most Help or Caused the Most
             Difficulty in Adapting to New Machinery 66

Table 4-4:   Employment Effects of Machine Change 68

Table 4-5:   Changes in Income, Seniority Rights, and Fringe
             Benefits at the Time of Machine Change 71

CHAPTER 5

Table 5-1:   Workers' Income and the Automation Level of Equipment
             Operated Directly and Used Indirectly. 78

Table 5-2:   Promotions, and Past and Expected Income Change and
             the Automation Level of Equipment Operated Directly . . .80

Table 5-3:   Unemployment and Workers' Relation to Equipment 82

Table 5-4:   Unemployment and the Automation Level of Equipment
             Operated Directly 82

Table 5-5:   Machine Change and Income .84

Table 5-6:   Promotions, Past and Expected Income Change and
             Machine Change 87

Table 5-7:   Distribution of Workers by Unemployment and Machine
             Change 89

Table 5-8:   Workers' Perceptions of Steadiness and Advancement
             in Relation to Machine Change .92

Table 5-9:   The Relation of Socio-Economic Variables, Machine Use
             and Machine Change to the Incidence of Past Income
             Increases 97

Table 5-10: The Relation of Socio-Economic Variables and Machine
             Change to the Seriousness of Past Unemployment 102

CHAPTER 6

Table 6-1: Change in Job Satisfaction over the Past 5 Years
            among Job and Machine Change Groups 111
Table 6-2: Relationship of Basic Demographic and Economic
            Variables to Job Satisfaction 113
Table 6-3: Importance of Relationship of Some Additional
            Demographic and Economic Variables to Job Satisfaction . 115
Table 6-4: Correlation of 16 Job Characteristics with Job
            Satisfaction and Income Change 116
Table 6-5: Importance of Relationship of Job Characteristics
            to Job Satisfaction 118
Table 6-6: Changes in Job Characteristics among Job and Machine
            Change Groups 120
Table 6-7: Relationship of Job Demands Scale to Change in
            Satisfaction and Interest 127
Table 6-8: Relationship of Job Demands to Job Satisfaction 127
Table 6-9: Relation Between Change in Job Satisfaction and in Various
            Job Characteristics Ranked by Size of the Simple Correlation
            Coefficient for Education and Age Subgroups 128
Table 6-10: Major Socio-Economic Characteristics of Job Demands—
            Satisfaction Groups 130
Table 6-11: Correlation Matrix of Change in Satisfaction and Change
            in Job Characteristics 133

CHAPTER 7

Table 7-1: Distribution of Workers According to Adjustment
            Scales A and B 138
Table 7-2: Relationship of Basic Demographic and Economic
            Variables to Adjustment Scale A 140
Table 7-3: Adjustment Scale A Related to Industry 141
Table 7-4: Adjustment Scale A Related to Circumstances Under
            Which Workers Left Former Employer 143
Table 7-5: Number of Employers in Past 5 Years by Adjustment
            Scale A 143
Table 7-6: Relationship Between Automation Level of Equipment
            and Adjustment Scale A 145
Table 7-7: Relationship of Job Demands Scale to Adjustment Scales . 146

CHAPTER 8

Table 8-1: Distribution of Workers by the Automation Level of
            Their Equipment and Education 162

Table 8-2: Frequency of Vocational Education among Workers with
            Different Amounts of Formal Education 162
Table 8-3: Vocational Training by the Automation Level of
            Workers' Equipment 165
Table 8-4: Distribution of Courses Taken by White-Collar and Blue-
            Collar Workers Operating Various Types of Equipment 167
Table 8-5: Education and Workers' Attitudes Toward Their Job
            and Equipment 169
Table 8-6: Perceived Need for Additional Education and Perceived
            Use of Skills and Training Related to Machine Change
            and Job Change 173
Table 8-7: Automation Level Related to Perceived Need for Additional
            Education, and Perceived Use of Skills and Training 174
Table 8-8: Relation of Vocational Education to Adjustment Scale A . 176
Table 8-9: Relation Between Education and Job Demands 177

APPENDIX II
Table II-1: Approximate Sampling Errors of Percentages 235
Table II-2: Approximate Sampling Errors of Differences 236
Table II-3: Percentage Distribution of Labor Force by Age, 1967 . . . 237
Table II-4: Percentage Distribution of Labor Force by Years of
             School Completed, 1967 238
Table II-5: Percentage Distribution of Labor Force by Sex, 1967 . . . 238

APPENDIX III
Chart I: Automation Level Classification of Equipment 244

Chapter 1

INTRODUCTION

The impact of technological change on the work force has been much discussed; it has been written about extensively. Yet it remains a controversial subject. Clearly, the work force as a whole benefits from technological advance which raises the productivity of labor, facilitates wage increases, permits more leisure, and sometimes may lead to price decreases. Rising incomes and occasionally falling prices widen the market for consumer goods and for the capital goods which produce them, thereby creating employment opportunities. In the long run and for the economy as a whole, changes in machine technology are no doubt a major source of growth of income as well as of employment. Yet some groups in the labor force may suffer, and these groups are of concern to government policy and programs. In particular instances technological advance may replace men by machines or may reduce the machinery *and* men needed to produce a given volume of output. Hence technical progress may lead to unemployment in the short run. It also may make human skills obsolete and thus reduce the earnings of some workers. Finally the automated factory or office may be a less satisfactory place to work than the more traditional work setting.

Thus, technological advance can have benefits as well as drawbacks for the worker.[1] Economic analysts for the most part would agree that the benefits are greater; but they have not been able to measure both of the opposing effects and to balance them against each other. It is the purpose of this study

---

[1]The expression "worker" is used in this study to refer to all members of the labor force, *not* blue-collar workers alone.

to provide some quantitative empirical evidence on the positive as well as negative consequences of changes in machine technology for the labor force as a whole. And a further objective is to determine under what kinds of circumstances adverse effects are most likely to occur.

To date, analytical work in this field by economists, to the extent that it was focused on the economy or labor force as a whole, has been largely theoretical.[2] It has examined the consequences of technological change which are likely on a priori grounds. But lack of appropriate data has made it impossible to put these deductions to quantitative empirical tests. To be sure, excellent data on unemployment, labor turnover, changes in earnings, occupational shifts, and geographic mobility are available for representative samples of the labor force. However, as yet we have not been able to determine precisely how these phenomena are linked to changes in machine technology.

Much of the empirical research on the economic impact of automation on the work force has consisted of case studies conducted in individual plants, companies, or narrowly defined industries.[3] One group of case studies has examined more or less controlled experiments which were conducted to evaluate management and personnel techniques that may be used to accomplish the transition. Another group of case studies, more closely related to the aims of this study, has traced the kinds of changes in occupational classifications, job characteristics, pay rates, or other labor force adjustments which accompanied a particular change in machine technology. Intensive studies of specific situations can contribute greatly to our knowledge, but they are not sufficient by themselves. The present study is the first empirical attempt to assess the impact of technological change on a *cross-section of the U.S. labor force.* As such it should be an important complement to the case study approach as well as to theoretical studies.

The survey on which this study is based was conducted in 1967 and interviewed a sample of 2,662 labor force participants—1,800 men and the rest women. These people lived in cities, towns and rural areas all over the United States; they worked in all kinds of industries and occupations. The sample was selected by multistage probability sampling techniques and is representative of the U.S. labor force. Selected respondents were interviewed an hour or more in their homes. A copy of the questionnaire is reproduced in Appendix I. Questionnaire design and interviewing techniques also are described in Appendix I. Sampling methods and sampling errors are discussed in some detail in Appendix II.

---

[2]For a review of this literature see Robert L. Heilbroner, "The Impact of Technology: The Historic Debate," in The American Assembly, John T. Dunlop, ed., *Automation and Technological Change,* Prentice-Hall, 1962, pp. 7-25.

[3]Some of the best of these studies are summarized in the U.S. Department of Labor, Bureau of Labor Statistics, Bulletin 1287, *Impact of Automation,* 1960, Part III; also Floyd C. Mann, "Psychological and Organizational Impacts," American Assembly, *op. cit.,* pp. 43-65.

The survey began by asking people in all occupations and industries to describe any machinery or equipment which they operate and to identify any other machinery which is important in connection with their work. Then members of the labor force were asked to report changes during the previous 5 years in any of the machinery they work with. The study first analyzes the extent of machinery use by the labor force as a whole and by various sub-groups and examines people's attitudes toward the machinery they work with (Chapter 2). It then measures the frequency with which jobs have been significantly affected by *changes* in machine technology and reveals what kinds of people have been most likely to experience such changes (Chapter 3). It also looks at the process of change-over to the new equipment—how much advance information people were given about the change, their participation in planning the change, how they were trained to work with the new equipment, how often the equipment change entailed upgrading of their position on the one hand or down-grading on the other, and how much unemployment occurred during the transition (Chapter 4). This inquiry is particularly concerned with the economic consequences of changes in machine technology on people's earnings, career advancement or setbacks, unemployment experience, job and geographic mobility, as well as their expectations for the future (Chapter 5). It is also concerned with the impact of technical change on perceived job characteristics and various aspects of job satisfaction (Chapter 6). The differences between workers who make a poor adjustment to technological change and those who make a successful adjustment are examined more thoroughly in Chapter 7. Finally, Chapter 8 is devoted to the role which education plays in the adaptation of the labor force to the advancing technology.

A. Defining Change in Machine Technology

Changes in machine technology range from those which are trivial, or at least peripheral to the operation of the machine and the worker's job, to those which constitute a veritable revolution in the factory or in the office. There is, in other words, a continuum of changes from the very marginal modification of a machine to the introduction of an entire automated system or line. There is also a continuum of changes which are mechanically simple to those which are highly complex from a technological point of view. Job content may also be altered to varying degrees. What then is the scope of this study?

The study might have been limited to technological changes which qualify as "automation" in a narrow sense. Strictly speaking, automation includes two kinds of changes: (a) automatic and integrated materials handling, conveying and processing equipment and (b) automatic control systems, incorporating "feedback" devices which reduce or eliminate the need for human control. Automatic control systems may pertain to machines which process things (solids, liquids, gases) as well as to equipment which processes information. There were several reasons for *not* confining the study to automation in this limited sense. It would have been difficult in the interviewing situation to draw a meaningful line between automation and other technological advances, such as various degrees of mechanization. Moreover, changes which strictly qualify as "automation" would have been experienced infrequently. And third, as far as individual workers are concerned simpler technical changes often have an impact which is comparable to that of automation. For example, a bus driver in Newark reported two moderate improvements in technology: the installation of automatic transmissions and change boxes (to collect the fare) on the buses. This change enabled the company to cut the work force on each bus from two to one and considerably affected the driver's work.

At the opposite extreme, a study such as this might have a very broad scope, dealing with all scientific advances which increase labor productivity. For example, in addition to the use of improved machinery, workers may become more productive when a plant is lighted better or air-conditioned, when they start to use an improved chemical process or (in the case of farmers) an improved fertilizer or pesticide, when the product is redesigned, when the work is organized more efficiently, or when management or personnel techniques are rationalized. A very comprehensive definition would have had the advantage of yielding a large number of cases in a sample of a given size who experienced a change during the reference period. However, it would have meant studying a very diffuse mixture of changes and experiences. The more broadly defined the phenomenon under study, the more difficult it becomes to construct concise and meaningful sequences of questions concerning it.

It appeared therefore that an intermediate concept would be most suitable for this study. Accordingly, the survey was focused on all changes in *machine* technology, but disregards other new applications of science to the production system. The survey thus examines not only instances where a machine took over all or part of a task formerly performed by human labor, but also the introduction of machines which perform a greater volume of work; which are faster, more accurate, more powerful, or more self-regulating; or which do higher quality work or put out goods and services previously not produced. In all these cases there is a presumption that job content will change to a greater or lesser degree concomitant to a *change in machinery.*

Needless to say, the study deals with people's perceptions of the impact of changes in machine technology on their job. These are bound to differ from the reports an engineer would have made, if he could have gone to the respondent's place of work and studied the change in question.

People's work is affected not only by changes in the machinery which they operate but also by machinery operated by others. For example, people who do maintenance and repair work operate power tools and other relatively simple equipment. In addition they are concerned with the complex factory equipment, or trucks, airplanes, etc. which they service. To mention one very different example, many people nowadays use computer output or have occupations which support the computer operation (keypunching, programming etc.). Many people's work has been greatly affected by the introduction of computers, yet few people operate computers. An assistant credit manager reported that he was literally replaced by a computer, which was not even located in his own department. It was decided that a study of changes in machine technology aiming to encompass the labor force as a whole should *not* be limited to the impact of such changes on machine operators, but should concern itself with all people whose work is significantly altered.

B. The Reference Period

At the outset it was necessary to decide how long a period of time the study should cover. In the interest of accurate recall and vivid reporting of the respondent's experiences a short reference period appeared preferable, such as asking people only about changes in machine technology that occurred on their job during the previous year or two. On the other hand, the shorter the period under investigation, the smaller would have been the proportion of the sample who could report a relevant experience. In the first pretest of the questionnaire a reference period of 3 years was used on an experimental basis. It became clear that changes in the equipment on the job are important events in people's working lives. Most people were eager and interested to talk about such experiences and seemed to remember them well. Indeed some respondents were disappointed that the interviewer did not want to hear about events which occurred more than 3 years ago. It was decided on the basis of the pretest experience that a reference period of 5 years was feasible. That is, it appeared that the gain in the number of cases would outweigh some possible loss of reporting accuracy.

Accordingly, this study covers the 5 years, 1962-67. This was a period of strikingly large growth of industrial production (34%) and real GNP (26%)[4] with relatively low levels of unemployment. The title of this study, *Technological Advance in an Expanding Economy,* is intended to remind the

[4]*Economic Report of the President,* February 1968, pp. 210, 249.

reader of the favorable economic setting to which the findings relate. That the study refers to a period of prosperity is, if anything, an advantage. For the installation of more modern equipment is accelerated when sales are rising, profits are high, when labor shortages make themselves felt, or labor costs are rising. Indeed automation and a growing volume of business can hardly be thought of as separate phenomena.

C. Advantages of the Cross-Sectional Approach

Much can be learned from a cross-section study of changes in machine technology which could not be learned by interviews in one or a few plants or from an analysis of company records. Most important, only a study of a representative cross-section of the labor force can show how frequent certain occurrences are, such as: experiencing *any* change in machine technology, being displaced by a machine and being unemployed for a time, being downgraded to a less demanding job, being upgraded, receiving a raise when new machinery is introduced, being retrained, and the like. And a cross-section study enables the researcher to go one step further and to determine the frequency of such events within occupation groups, education groups, and the like. For example, we shall see whether unfavorable consequences increase with age or decrease with age (because of seniority).

Only a cross-section study can yield a representative picture. The companies which permit a team of social scientists or even a government agency to study a change in machine technology in one of their plants may well rate above average in degree of planning and concern for the work force. A plant manager who was interviewed in connection with this study admitted that he gave his workers no notice at all when a part of the operation was automated. Why? "I didn't want those union fellows to tell me what to do." It is unlikely that a case study would be permitted in such a setting. Case studies may also be nonrepresentative in that they tend to focus on conspicuous one-shot changes in technology. Social scientists are likely to select for study technological changes which require relatively large work force adjustments within a short span of time, since such situations are particularly interesting and amenable to investigation. However, such situations are not representative. To quote from one of the studies prepared for the National Commission on Technology, Automation, and Economic Progress:

> "For the most part, technological innovations are not adopted extensively in an industry or in an individual plant at any single time. Instead they are often adopted piecemeal in the form of a great many minor

changes introduced in one establishment and then in another and often in a gradual way within an establishment."[5]

Interviews with automobile workers, for example, underline this point. Also, it seems fair to say that case studies have hardly ever been conducted in plants with obsolete facilities which had to reduce hours or employment because of their deteriorating competitive position.

The human factor is crucial when one considers the impact of technological change on the labor force. Studies at the micro-level, based on detailed interviews with a representative sample of workers, are most suitable for revealing economic and personal experiences as well as psychological reactions to work changes. For example, case studies which utilize largely company records can reveal only that John Smith, a skilled glass maker, was laid off; they cannot tell us what happened to him afterwards. Personal interviews about experiences in the past 5 years have a longer time perspective. They can tell us that of the people who were replaced by a machine and laid off in the past 2 to 5 years, X percent found another job at the same or a better rate of pay within 1 month; Y percent are now engaged in work which they consider less skilled or which they dislike, and so forth.

Often personal interviews will give more insight than company records. One example is sufficient to illustrate this point: A skilled carpet weaver who was interviewed in this study reported that he quit his job, when the other firms in the industry introduced automatic weaving. Why did he quit? "I saw the handwriting on the wall. We were working fewer and fewer hours." This man was in an economic sense replaced by a machine; he is now wrapping muffins on an assembly line, and regrets the obsolescence of his skills. Yet company records would show him as having quit voluntarily, without any changes in machine technology in the firm for which he was working.

D. Problems and Limitations of the Cross-Sectional Approach

Despite the important information which can be gained from a study of the labor force as a whole, there were good reasons for tackling the automation problem by case studies first. Case studies are much simpler. The cross-sectional approach involves difficult measurement problems. There is almost infinite diversity in the kinds of machines people use on their jobs and in the relation between the machinery and the worker. And there is also much diversity in the kinds of machine changes which may occur.   The owner of a

---

[5]Bureau of Labor Statistics, U.S. Department of Labor, "Industrial and Occupational Manpower Requirements," in *Studies Prepared for the National Commission on Technology, Automation, and Economic Progress,* Appendix Volume I, p. 177.

greenhouse may install a sprinkler system; a professional photographer may switch to a camera with a built-in lightmeter; a secretary may convert from a mechanical to an electric typewriter; a production line may be automated. This last change will have a different impact on the men on the line, the repair and maintenance crews, the supervisors, or the personnel and training department. To devise a series of questions which would be equally applicable to such diverse occupations and work changes *and* which would yield meaningful data was a complex task. We took up this task knowing that the first experiment would yield solutions that would not be entirely satisfactory.

People are for the most part well informed about the equipment they work with. Some delighted in describing it lucidly and in great detail; most others were able to give the interviewer at least the name of the equipment and to reply to a series of questions about its major functions and characteristics. A classification of presently used equipment by degree of automation was developed and is used in the analysis which follows. The classification and the conceptual scheme underlying it are described in detail in Appendix III. The questions on the respondents' equipment were coded by graduate students trained in engineering.

People's accounts of the nature of the technological *change* which they experienced were less comprehensible than their descriptions of the equipment they use. The questionnaire seems to have been effective in ascertaining the presence or absence of equipment changes. It was not feasible, however, to measure or classify these equipment *changes* by their technical characteristics, since respondents were often unable to convey this kind of detailed technical information. This is not a serious shortcoming of the study. Although the starting point of our analysis is equipment change, our interest is focused on the resulting changes in work, income, employment, job content, and the like. People could describe these without difficulty.

One of the most important and controversial aspects of machinery change is its impact on labor demand. Obviously, this study can make only a limited contribution to that particular aspect of automation research. For example, it does not throw any light on the question of how many new jobs are created in the machinery industry by the advancing technology. Nor can it tell us to what extent advances in machine technology have added to the demand for other goods and thereby created more jobs.

To appreciate what this study can and cannot contribute to the analysis of technological change, one must distinguish between the direct and the indirect consequences of technological advance. For example, the raises in wages received by workers whose equipment is modernized may be viewed as a *direct* consequence of technological change. Widely diffused income increases to other workers, facilitated by the growing productivity of the economy, are *indirect* consequences of technical advance.

The same distinction must be made with regard to unemployment. In the terminology of this study, *direct* unemployment is that which is experienced by the men who used to work with the equipment which was modernized *and* who could not immediately find another job. It also would include the unemployment of the low-seniority worker in the same firm who might be dismissed instead. *Indirect* technological unemployment may hit workers in firms which cannot compete with their technologically more advanced competitors. Indirect unemployment also is experienced by workers who might have been hired (and hence would have suffered less unemployment), if labor saving machines had not reduced the demand for their kind of labor. Indirect unemployment is too diffuse and too hard to trace to lend itself to survey measurement, just as is true of the employment-creating effects of technological advance.

The distinction between direct and indirect unemployment is vital for policy purposes, because it distinguishes between rather different kinds of unemployed people. Direct unemployment affects people who have experience and know-how in a line of work and usually a satisfactory work history. Direct unemployment is a kind of structural unemployment. Indirect unemployment is experienced by people who might have been hired, if there had not been so much mechanization and automation, i.e., those at the back of the line in the competition for jobs. These may include many new entrants into the labor force, the old, the unskilled, the handicapped, and other disadvantaged or unsuccessful groups.

This study measures the volume of *direct* unemployment occasioned by technological change. It yields an approximate estimate of the percentage of people whose job was abolished by new machinery in the past 5 years, the percentage who became unemployed for various periods of time during the transition, the percentage who were immediately reabsorbed. It also compares the unemployment experience (irrespective of cause) of people who work with various kinds of machines with the unemployment experience of those who do not use equipment. And finally it compares unemployment in the past 5 years among all those who experienced technological change and all those whose equipment remained the same.

The unemployment estimates are meaningful only in the context of a particular economic environment. If the economy had been more stagnant than it was in 1962-67, the same changes in machine technology might have displaced more workers. Some employees reported that their firms experienced greater sales after the new equipment was installed and hence needed the same number, or even more workers, in spite of the fact that the new equipment was faster, more productive, or labor saving. We cannot expect the average respondent to disentangle a potential decline in employment in his firm due to new machinery from offsetting increases which are due to the growth of aggregate demand or other improvements made by the firm, more or less independent of the technological change. Even the employer would often be unable to make such a breakdown. In short, the present study enables us to estimate the frequency with which unemployment, downgrading of skills, pay cuts or income increases were *associated* with changes in machine technology, *given the prevailing economic conditions.*

These limitations of the study should be understood, but they should not be exaggerated. The study relies on the reports of workers to determine whether unemployment occurred in connection with the change in machine technology and whether job changes, transfers, and layoffs were brought about directly by such technological change. Practically all members of the labor force are keenly interested in the reasons for their job progress or for adverse experiences. Workers discuss among themselves mechanization, automation and the progress of their firm, and generally their impressions are realistic. For example, some workers reported that they were transferred because their old job was taken over by a machine, and that workers with less seniority in another department (whose jobs were not eliminated) were laid off.

E. Some Major Conclusions

Over a 5-year period about 10 percent of the labor force underwent one or more changes in machine technology which (in their own view) altered their work significantly. Another 12 percent experienced a machine change as a result of a job change, where the job change was (again in their own view) not caused by a change in machine technology. True, in some cases workers may not see some of the more remote and indirect forces which make a job unsatisfactory and induce them to seek other employment. Still, we must conclude that, percentagewise, technological advance changed relatively few jobs to a significant degree—about 2 to 3 percent a year. In rough absolute terms such percentages imply 1.5 to 2 million members of the labor force a year, not a small number at all.

The rather limited impact of technological change on the labor force, percentagewise, is not difficult to explain. There is no question that technological advance affects a much larger proportion of the total stock of machines than of the total work force. For one thing, some extensive machine changes hardly alter the jobs of the people who work with them or affect the work of only a small proportion. Secondly, highly automated or mechanized equipment requires relatively little labor to tend it. The largest synthetic ammonia plant in the world, built in 1965 by Olin-Mathieson Chemical Corporation, was designed to operate with a grand total of 32 employees, 11 of them supervisory, technical and clerical personnel. A steam-electric plant now in operation requires a single worker per shift. Thus jobs related to the "new" machines are created in limited numbers, while the economic plenty which automation makes possible expands employment opportunities in many traditional kinds of work.

Most commonly newly mechanized or automated jobs are seen as requiring experience, skill and personal capabilities. They therefore are staffed by workers recruited within the same department or at least the same company. In addition, there has been some tendency for young workers to obtain jobs requiring the use of the newer equipment, while older workers have tended to retire from jobs involving more traditional equipment. There is no evidence that shifts from one employer to another were a major mechanism by which jobs which relate to newly introduced equipment were filled. Of course, many job shifts involve a movement upward on the Automation Scale—to technologically more sophisticated equipment; but a similar number entail a shift to equipment at a lower level of technology. And the largest proportion of job changes entail no significant equipment change at all.

The Report of the National Commission on Technology, Automation, and Economic Progress suggested that:

> "Technological change. . .has been a major factor in the displacement and unemployment of particular workers. Thus technological change (along with other forms of economic change) is an important determinant of the precise places, industries, and people affected by unemployment. But the general level of demand for goods and services is by far the most important factor determining how many are affected, how long they stay unemployed, and how hard it is for new entrants into the labor market to find jobs."[6]

The survey findings do not support the first part of the Commission's statement. There is little evidence in the data that technological change determines the precise incidence of unemployment. Rather the survey leads us to infer that the impact of advances in machine technology on employment is largely indirect. The firm which introduces new labor-saving machinery often can rely on normal retirements and resignations plus an expanding market to bring its labor supply in balance with needs. Skilled and experienced people who have to be laid off tend to be re-employed quickly. Much of the unemployment resulting from labor-saving machinery "trickles down" to the most marginal groups in the labor force. Workers who might have been hired in the absence of technological change are not needed. The last to be hired

---

[6]National Commission on Technology, Automation, and Economic Progress, *Technology and the American Economy,* Vol. I, U.S. Government Printing Office, 1966, p. 9.

have to wait longer for a job; they suffer more unemployment. The last to be hired are the people with the weakest labor market qualifications for reasons of age, experience, education, health, location, race, previous employment record and other possible handicaps. These people, together with those employed by technologically backward firms, seem to bear the bulk of technological unemployment. True, some workers are laid off when new machines take their place; but a large proportion of those who remain out of work for any length of time have a weak competitive position in the labor market to start with. Again some were poorly educated; others were elderly, Negroes, residents of small towns, working wives who were very limited in their working hours and geographic location, and the like. Although a large proportion of the reported machine changes were labor saving, their employment impact was cushioned by the fact that improved machinery typically is acquired by the firm faced with an expanding market or with a shortage of qualified labor. Indeed the survey suggests that technological unemployment often occurs in firms which are technologically backward and hence noncompetitive. As such firms decline, they tend to offer relatively poor pay, short hours, and unsteady work. The workers who remain with such employers are those who by reason of their weak labor market qualifications have the most difficulty finding alternative employment. When they finally decide to or are forced to change jobs, they may have to go through a period of unemployment before they manage to find other work.

The survey findings *do* support the second part of the National Commission's statement which contends that aggregate demand is *the* key to the economy's ability to absorb technological unemployment. There is no need to defend or discuss at length the proposition that the answer to the job problem of the marginal worker is a strong demand for labor throughout the economy plus job training, with the major emphasis on labor demand. A strong demand for labor derives from a strong demand for goods and services, which in turn can be stimulated by fiscal and to some extent monetary policy. It may be concluded from the survey data that in retraining and other special programs to ease the impact of technological unemployment *the marginal segments of the labor force* should receive particular attention. To be sure, there are instances where skilled workers with strong employment qualifications need retraining. Private employers in particular should be encouraged to extend existing programs and devise additional ones which will help to fit these workers into new jobs.

Although the skilled and experienced worker who has been displaced by a machine and cannot find another job within a short time is an uncommon phenomenon, one may still suspect that such workers would suffer income losses as a result of technological advance. After all, if the demand for certain kinds of skilled labor is reduced, its ability to secure income advances may also be reduced. The survey indicates that any such tendency is outweighed by the greater productivity of the new equipment. Many more workers receive income increases during the transition than suffer pay cuts, although overtime hours are often reduced. Over a 5-year period workers who experience a change in machine technology receive income gains more frequently than workers who continue to work with the same equipment; this is true despite the occasional unemployment which is associated with machine change. Those who use the technologically most advanced equipment likewise fare better in terms of income level and income change than those who work with more traditional equipment. A large part of the observed favorable income differentials associated with highly automated or mechanized equipment is attributable to the above-average educational qualifications of those who work with such equipment and those who experience changes in machine technology. After allowance is made for educational and other socio-economic differences, a small income advantage remains for those whose jobs have undergone a change in technology. It would appear then that the well-educated derive disproportionate benefits from technological advance; yet nobody would seriously suggest that they are the only beneficiaries. The small size of the net income differential in favor of users of technologically advanced equipment suggests that the larger part of income increases made possible by productivity gains are widely and promptly diffused through the economy. At any given level of education those who are not directly affected by changes in machine technology benefit to almost the same extent as those who are.

Writings on automation by intellectuals have often focused on the helpless "victims" of automation or have implied that we are all on the way to futility. Occasionally union officials and groups of workers fearing displacement by new machines have joined in criticizing our rapidly advancing technology.

". . .we can. . .by no means justify the naive assumption that the faster we rush ahead to employ the new powers for action which are opened to us, the better it will be. . ." (Norbert Weiner, "Some Moral and Technical Consequences of Automation," in Morris Philipson ed., *Automation, Implications for the Future,* 1962.)

"The belief in the benign social impact of technology may turn out to have been the most tragic of all contemporary faiths." (Robert Heilbroner, "The Future of Capitalism," *Commentary,* April 1966.)

Interestingly, both these statements were quoted in an AFL-CIO publication.[7]

The survey indicates that the large majority of U.S. labor force members do not share such misgivings. For the most part Americans value the

---

[7]AFL-CIO, *Labor Looks at Automation,* Washington, D. C., 1966, p. 11.

equipment they work with, and believe that automation is a good thing for people in their line of work. Those who work with highly automated or mechanized equipment are more likely than others to say that they enjoy their work. People who have experienced a recent advance in machine technology express increased satisfaction with and interest in their job with greater frequency than those who have worked with the same equipment for 5 years. It was necessary to suggest specific disadvantages of a technologically advanced work environment to respondents in order to elicit even some marginal indications of dissatisfaction from those who are part of such an environment. If anything, positive evaluations of automation are more pronounced among the better educated workers than among others. Yet a major survey finding is that these favorable attitudes are pervasive in all U.S. population groups. Needless to say, this particular finding may not apply equally to other countries. The American labor force is receptive to change. We know relatively little about the conditions which determine attitudes toward technological change in other countries.

The reasons for the low frequency of negative reactions to automation on the part of the American work force are not purely economic. Of great importance is the fact that changes in job content and job characteristics associated with technological advance for the most part have favorable connotations. The survey indicates that people who experienced changes in machine technology tend to view their jobs as having become more demanding. They note frequently that the new machinery is faster than the old and thus calls for more attention, alertness, and a reorganization of the work process. They are likely to report that new skills must be acquired, that there is more need for planning and judgment, that there is more chance to learn new things, that errors have become more serious, that they must take more initiative in organizing their work. Increased job demands in turn have an important and positive impact on job satisfaction and job interest in all major socio-economic groups. A number of case studies found "job enlargement" to be a consequence of automation, but other case studies found greater monotony and a downgrading of skills. The survey reveals that the first situation is much more typical than the second.

There are other changes besides. According to the people affected, advances in machine technology very often reduce physical work effort. Sometimes they are accompanied by an improvement in the work environment—less heat and noise, better lighting and ventilation. Greater opportunity to talk to fellow workers was reported more often than increased physical isolation. On balance advances in machine technology are also seen as reducing the danger of personal injury. Interestingly, none of these environmental consequences of technological change are as conducive to greater job satisfaction as are reports of greater job challenge or job demands. No doubt there are conditions under which changes for the better in the job setting and physical work requirements are as crucial, or even more crucial, for job satisfaction than is greater job challenge. However, seriously objectionable work environments or physical work burdens may have become quite uncommon some years ago. An improvement in an already agreeable work situation cannot be expected to enhance job satisfaction to a great extent.

It is evident that as the educational attainments of the work force rise, an important task for management is to think about and devise methods of providing the job challenge and opportunities for personal growth which workers desire. In some cases job challenge results inadvertently from technological advance. The management's problem then is to design training programs and systems of supervision which will facilitate adjustment to the new machines. In other cases work challenge may have to be created deliberately through increased emphasis on "promotion from within"—upgrading from the lower echelons of the work force (there were repeated complaints in the survey about the practice of filling desirable new positions with "outsiders"), by making available more training and learning opportunities, by work rotation systems, or by transferring more responsibilities from supervisory to other personnel.

The survey confirms earlier evidence that the advancing technology requires a well-educated work force. There is a very pronounced relation between the technological level of equipment and the formal education of the people who work with this equipment. The more mechanized or automated the equipment, the higher the educational level of the work force associated with it. Moreover, the probability that a worker will experience an advance in machine technology is positively related to his education, particularly years of schooling. However, the rate at which educational requirements are rising may not be as rapid as is sometimes supposed (after all, the survey indicates that over a 5-year period the work of a majority of people was not greatly altered by technological change).

Whether the high level of education associated with modern equipment is really needed or whether employers set unnecessarily high educational qualifications for people who are being hired to work with the latest technological systems cannot be determined on the basis of this survey. The survey shows that the well-educated people who work with the most mechanized and automated equipment do not complain more often than others that they have unused skills. Indeed, they frequently express a felt need for further education and training. The modern technology obviously creates some new demands for staff with scientific or advanced technical training. There is no convincing evidence in the survey that the personal capabilities which help a worker to meet the enlarged job demands of the new equipment constitute an additional link between technologically advanced machines and the highly educated work force associated with such machines.

Chapter 2

MACHINERY USE BY THE WORK FORCE

In the American economy many people work with machines. Some members of the labor force operate machines; others install, clean, repair, sell, or design machines. Some workers train people to use machines, or supervise those who operate machines. Still others provide input for machines or are dependent on some kind of machine output in connection with their work; but beyond this the picture of the worker and the machine has never been brought into sharp focus. No previous attempt has been made to estimate the proportions of the U.S. work force which relate to machines in various ways or the proportion which does not use machines at all. We have not known how continuously or intensively jobs involve the use of machines. We also have not known how many people use computers and highly automated equipment on the one hand, and how many still use simple hand tools on the other.

Section A of this chapter examines the ways in which members of the U.S. labor force work with the machinery on their jobs and the extent of its use. It also estimates the frequency with which various kinds of equipment are used, including a rough estimate of the level of automation of the equipment. Further, the chapter will relate the use of machinery to socio-economic characteristics of workers such as age, education, race, occupation and industry, and size of the firm for which they work (Section B). Another consideration, at least as interesting as people's use of machinery, is their attitudes toward machinery. From an economic point of view machines may help the worker to be more productive, give him more leisure, and enhance his earnings. From a social-psychological standpoint machines are sometimes described as dehumanizing, preventing the worker from organizing his own work, talking, or making his own decisions, turning him into a robot who must respond invariably to the uncompromising demands of the machine. The movie *Modern Times* probably has been *the* classical expression of this view. Section C will look at the ways in which members of the labor force react toward machines and automation and how these attitudes are related to their use of machinery.

The focus of this monograph is on *change* in machine technology. In the interviews a detailed series of questions on present machinery *use* served as an introduction to the inquiry about machinery *change.* It led the respondent to think of the various kinds of machinery that have a bearing on his job. Thus the questions concerned with change were asked after a broad frame of reference had been established. Yet the information on frequency of machinery *use* also is of interest in its own right. And in addition, an analysis of machinery use and the reactions of the worker to machinery use is indispensable for an understanding of the impact of change, to which we will turn in the following chapter.

A. The Frequency of Machinery Use

The most basic distinction with regard to machinery use is that between people who operate machines and those who do not. A further distinction may then be made between people who operate machinery all day long (many production workers), those who operate it some of the time (a secretary or a dentist), and those who operate it very little (an economist with a calculating machine). A classification of the labor force along these dimensions is presented in Table 2-1. The exact questions on which the classification is based appear at the bottom of the table. The table shows that for a large segment of the American labor force machine operation is an important aspect of their work. About two out of every three labor force members operate machinery or equipment in connection with their work.[1] The data relate to people in all occupations and industries including business executives, lawyers, and teachers as well as factory workers, truck drivers, and waitresses. Still, only 28 percent of the American work force reported that they operate no machinery at all.

---

[1]About 2 percent of the sample reported operating *only* hand tools. The decision to count these people as equipment users is debatable. It is, however, inconsequential because of the small proportion of cases concerned.

TABLE 2-1

PROPORTION OF MEMBERS OF THE LABOR FORCE WHO OPERATE MACHINERY
OR EQUIPMENT ON THEIR JOB, BY SEX

Operation of equipment	All	Men	Women
Almost constantly	42%	40%	47%
Some of the time	20	20	20
Very little	9	9	8
Not at all	28	30	23
Not ascertained	1	1	2
Total	100%	100%	100%
Number of cases	(2,662)	(1,800)	(858)

The questions asked were: "On your present job, do you ever operate or help
to run any kind of a machine, mechanical equipment, or power tools on your
job? I mean tools or machinery that have a motor or a gasoline engine, or
equipment that is run by electricity or air pressure, or anything like that.
What kinds of equipment or machinery are these? Do you use this almost constantly, some of the time, or very little?"

Interestingly, more of the employed women than men operate machinery. As
many as 42 percent of workers said that they operate machinery almost constantly.[2]

The degree of involvement with machinery appears even more striking
when one considers that a good many workers' jobs center around machines
which they do not operate. The airplane is a good case in point: Only one or
two men are needed to fly it, but repair and maintenance crews are numerous.
The computer is another example: Very few people operate it, but many
more people are needed for repair and servicing, programming, keypunching,
and other tasks in preparing the information input. A great many others
work with the data output. The manager of a bank which has recently had
many of its operations computerized said:

> "By and large, most people at the bank have only indirect connections
> to the automated equipment, but the effects of automation are fairly
> pervasive. There is no place where there is no connection whatsoever to
> the automated equipment."

One should perhaps think of the impact of a large piece of automated
equipment as analogous to a series of concentric circles. The operators usually
(but not always) are in the innermost circle and are affected most. Moving

[2] For the sampling error of the percentages in the text and tables of this monograph, see Appendix Table II-1. For the sampling error of differences between percentages, see Appendix Table II-2.

outward, in each consecutive circle the effects may be smaller but also may involve more people.

Since the relationship between workers and machines which they do *not* operate varies on a continuum from a very· close to a very peripheral involvement, the question arises how one should classify people with regard to this dimension of machinery use. It was decided to permit the respondent to classify himself (or herself). After a series of questions about equipment operated by the respondent himself, this further question was asked:

"Sometimes equipment which a person doesn't operate, or operates very little, can be important for his work. I mean equipment in your own department, in other departments, or even equipment in other companies. Is there any equipment which you don't operate which is important to you because a change in that equipment would affect how you do your work?"

At least a third of labor force members reported equipment which they do not operate but which they consider important for their work. A few cases from the survey serve to illustrate the variety of situations encompassed by this category:

A 31-year old worker stands at the end of a drive chain in a plywood factory. The drive chain carries the plywood through driers and to the man who unloads the plywood sheets and puts them on dollies. He has no control over the drive chain and operates no equipment himself.

The president of a private hospital described the extent to which he depends upon the medical and test equipment located in the hospital. Of course, he does not operate this equipment. The respondent plans on moving his hospital to a location near a large medical complex, in order to have access to better and more expensive outside equipment.

A senior mechanical engineer in the bottle cap division of a glass company designs and improves production equipment and modifies the design of equipment so that special orders can be filled. He never operates the equipment himself.

A traveling salesman for a wholesale electrical, plumbing and air conditioning firm has available to him an IBM 520 computer to calculate profit ratios and costs on large and complex orders. He feels it helps to determine if he is making a "fair" profit. The computer also handles billing and inventory.

An employee of the Library of Congress reviews and selects reports. She uses a computer to help in the selection. She merely makes requests for output from the computer; she never runs it nor does she even know where it is located.

A secretary in a public school depends upon keypunches and computers in the state department of education which keep track of teachers' records and relieve the secretaries of some of the former burden of filing and record keeping.

Needless to say, a worker who operates equipment may feel that some additional equipment which he does not operate is also important for his work. In the case of a man who operates a power tool on a production line, our conceptual scheme considers the power tool as equipment he operates and the conveyor or assembly line as other equipment which is important for his job. Nearly two-thirds of those who reported indirect use of machines[3] also indicated that they operate some other equipment. Table 2-2 shows the incidence of machinery use by members of the labor force, distinguishing between those who operate equipment, those who said equipment which they do not operate is important for their work, those who fall into both categories, and those whose jobs, according to their own reports, have no significant relation to machinery or equipment. The small group who operate equipment, but operate it very little (see Table 2-1), were classified in Table 2-2 and throughout the rest of this chapter as not working with machinery, unless they reported that other equipment was important for their work. Given this broader, and probably more meaningful, measure of machinery use, the proportion of people for whom machinery has an important bearing on their work appears to be between 75 and 80 percent of the labor force.[4]

TABLE 2-2

PROPORTION OF PEOPLE WHOSE WORK IS SIGNIFICANTLY AFFECTED
BY MACHINERY OR EQUIPMENT, BY SEX

Workers' relation to equipment	All	Men	Women
Both operate and have indirect use	20%	22%	17%
Operate only	42	38	51
Indirect use only	14	18	6
No significant machinery use	22	21	25
Not ascertained	2	1	1
Total	100%	100%	100%
Number of cases	(2,662)	(1,800)	(858)

[3]In the rest of this monograph, for the sake of stylistic simplicity, the expression "indirect use of machines" will often be employed to refer to cases where a worker regards machinery as important for his work but does not operate this particular machinery.

[4]In certain cases the coders were instructed to modify the respondents' answers in order to avoid meaningless or inconsistent classifications. For example, a telephone is

Continued on following page

There is of course almost infinite diversity of the kinds of machinery reported. When asked what kind of machinery they work with, the cross-section of labor force participants enumerated on the average three pieces of equipment they operate for each piece of equipment which has an important indirect effect on their work. Table 2-3, Part A, presents a very broad classification of all machinery mentioned. It distinguishes between machines which people reported operating (column 1) and machines which were reported to be important to their work, although operated by others (column 2).[5] The differences between columns 1 and 2 are pronounced. The more traditional and smaller scale equipment—whether it is production, office, or professional equipment—involves primarily direct contacts between the operator and the machine, although some indirect contacts were reported also. Computers and other electronic office equipment as well as large-scale production equipment on the other hand make up a relatively large proportion of machines which affect people's work indirectly.

A similar conclusion emerges from Table 2-3, Part B, which presents a distribution of labor force members by the technological level of their most sophisticated equipment. Machinery reported by respondents was classified according to its technical characteristics, with the aim of constructing a scale which would rank-order people from those working with the least (bottom) to those working with the most (top) automated kinds of equipment. The concepts underlying this "Automation Scale" are discussed in detail in Appendix III. The scale was coded by engineering students at the University of Michigan on the basis of technical information and the name of the machine, as supplied by respondents. The classification no doubt is sometimes inexact.[6] Yet an approximate rank-ordering of equipment has been achieved. We find that about 6 percent of people who *operate* equipment work with computers or

---

Footnote Continued

important in many work situations, but was disregarded except in the case of telephone operators, telephone repairmen, etc. Similarly, typewriters are important in many offices; the rule governing coding was that the fact that a businessman's or lawyer's secretary has a typewriter should never be sufficient to put him into the category of indirect machine use. Many people drive automobiles to work or on business trips. Only truck drivers, taxi drivers and the like were classified as operating a motor vehicle on their job.

[5]It should be noted that the upper part of Table 3 presents a distribution of equipment rather than of people. Each machine reported is one case — up to three cases for the same respondent. The lower part of Table 2-3 presents a distribution of people who work with machines.

[6]Although the engineering coders were encouraged to use the "Not ascertained" category liberally in case of doubt, they encountered only a small proportion of cases which they felt unable to classify (as indicated by the small proportion of "Not ascertained" cases in the lower part of Table 2-3). For a further discussion and an estimate of accuracy, see Appendix III.

TABLE 2-3

TYPE AND AUTOMATION LEVEL OF MACHINERY
IMPORTANT TO WORKERS' JOBS

A. Type of machinery	Machinery operated by workers[a]	Other machinery important to workers' jobs
Transportation vehicles - all types, passenger and freight	8%	9%
Small-scale office equipment	25	8
Tools, small-scale production, household, institutional equipment	29	14
Mobile equipment - construction, farm, freight	7	8
Computers, electronic equipment	6	27
Production equipment - large scale, non-mobile	14	30
Professional, specialized equipment	4	3
Miscellaneous equipment	7	1
Total	100%	100%

B. Automation level[b]		
Numerical, tape, or computer control	5%	21%
Other logical control	1	1
Fixed mechanical control	54	50
Powered multi-system, manual main control	21	20
Powered single-system, manual control	16	5
Operator powered and controlled	2	*
Not ascertained	1	3
Total	100%	100%

---

*Less than 0.5 percent.

[a]The term workers here includes all members of the labor force.

[b]People who worked with several machines were classified according to their most automated piece of equipment, unless this piece was one which they reported using "very little." Includes only those who work with equipment.

equipment controlled by tape, computer or other logical controls. These high automation categories account for 22 percent of the equipment with which people have indirect contact of some importance. Not surprisingly, computers require few operators, and even the control system of an automated factory is staffed by a small number of employees. An important part of the influence of automated equipment on the work of the labor force occurs in indirect ways, via preparation of inputs, transfer, coordination, use of output, maintenance, and the like.

B. Machinery Use in Relation to Socio-Economic Characteristics of the Labor Force

The very widespread use of machinery itself indicates that machinery must be important for almost all groups in the labor force, irrespective of occupation, industry, education, income, age, or size of employer. The data bear out this supposition, but they reveal some interesting differences also.

In every major occupation the proportion of workers who say that their job involves the use of machinery, directly or indirectly, is at least 60 percent. Also, in every major occupation the porportion who *operate* equipment at least some of the time is above 38 percent. Absence of significant machine use is most characteristic of professional and service workers and non-farm laborers (Table 2-4). People who operate machines are found with particular frequency among operatives, foremen and craftsmen, farmers, clerical workers and the self-employed. There appears to be some tendency for people in higher status occupations to report that what they do on their job is affected to an important extent by equipment which they do not operate. This situation was indicated by a rather high percentage of managers and officials, as well as the self-employed.

Among industries, professional services, government, and trade have the highest proportion of people whose work is entirely unrelated to machinery and equipment. Yet, even in these fields the proportion who use no equipment at all is at most a third of the work force. Manufacturing is *not* characterized by as high a concentration of people who operate machines as are construction, transportation, communication, and other public utilities, agriculture, and certain service industries. In fact, operation of machinery is so common in most industries that manufacturing does not stand out in this regard. However, in the goods producing industries many of those who operate equipment said that they operate it continuously, while in the service industries more people reported that they operate equipment only "some of the time." Since many units in manufacturing, transportation, and public utilities are large, there is a high degree of differentiation of functions among workers. As a result relatively many people report that machines operated by others are important for their work.

TABLE 2-4 (Sheet 1 of 2)

DEMOGRAPHIC CHARACTERISTICS OF PEOPLE WITH
DIFFERENT PATTERNS OF EQUIPMENT USE

	Workers' relation to equipment						
A. Occupation	Both operate and have indirect use	Operate only	Indirect use only	None	Not ascer- tained	All	No. of cases
Professional and technical workers	15%	32%	19%	33%	1%	100%	398
Managers, officials	14	24	33	29	*	100	185
Proprietors, self- employed businessmen	46	21	21	10	2	100	148
Clerical and sales workers	17	49	10	24	*	100	531
Craftsmen, foremen	32	36	19	12	1	100	421
Operatives	19	58	7	14	2	100	508
Service workers	6	48	6	36	4	100	222
Laborers - nonfarm	11	37	15	36	1	100	92
Farmers and farm workers	26	55	3	14	2	100	137
B. Industry							
Agriculture, forestry, fisheries, mining	25	51	5	16	3	100	166
Construction	24	42	13	18	3	100	180
Manufacturing	19	41	20	19	1	100	805
Transportation, communication, utilities	30	39	18	13	*	100	193
Trade - wholesale, retail	18	43	12	25	2	100	402
Finance, business services	24	39	13	23	1	100	196
Repair, personal and entertainment services	20	46	8	22	4	100	140
Health, education and welfare services	14	45	6	34	1	100	368
Government	21	35	17	25	2	100	174
C. Size of firm's employment							
Multiplant firm[a]							
Less than 50 employees	21	44	12	23	*	100	421
50-499	22	39	15	24	*	100	523
500-1,999	20	41	18	21	*	100	242
2,000-4,999	20	34	27	18	1	100	148
5,000 or more	13	35	27	24	1	100	176
Single plant firm							
Less than 50 employees	24	43	10	20	3	100	730
50-499	14	48	12	25	1	100	213
500 or more	18	47	9	26	*	100	68

TABLE 2-4 (Sheet 2 of 2)

DEMOGRAPHIC CHARACTERISTICS OF PEOPLE WITH
DIFFERENT PATTERNS OF EQUIPMENT USE

	Workers' relation to equipment						
D. 1966 income from worker's job	Both operate and have indirect use	Operate only	Indirect use only	None	Not ascertained	All	No. of cases
Less than $3,000	13%	49%	7%	28%	3%	100%	515
$3,000-4,999	18	54	5	21	2	100	555
$5,000-7,499	23	44	13	20	*	100	689
$7,500-9,999	25	36	18	21	*	100	396
$10,000-14,999	26	25	26	23	*	100	297
$15,000 or more	25	15	39	19	2	100	155
E. Education							
0-7 grades	15	47	5	29	4	100	212
8-11 grades	17	50	11	20	2	100	618
8-11 grades plus vocational training[b]	26	44	17	12	1	100	145
High school degree	20	47	11	21	1	100	568
High school degree plus vocational training[b]	27	44	14	14	1	100	286
Some college[c]	22	38	17	22	1	100	407
BA degree or higher degree	20	26	21	31	2	100	383
F. Age							
Under age 25	21	52	7	18	2	100	347
25-34	21	46	12	21	*	100	572
35-44	22	38	19	19	2	100	645
45-54	20	42	15	22	1	100	610
Age 55 or older	17	38	12	30	3	100	480
G. Race							
White	21	42	15	21	1	100	2365
Negro	11	49	6	30	4	100	229
Other	11	45	6	38	*	100	44
All workers[d]	23	45	11	17	4	100	2662

---

*Less than 0.5 percent.

[a]Number of employees in the respondent's plant or office.

[b]A vocational course of three months or more.

[c]One to three years of college, including all degrees below the BA level.

[d]Includes those who may have been "not ascertained" in any of the above tables.

This interpretation is corroborated in Part C of Table 2-4. Jobs for which equipment operated by others is important are found to a greater extent in medium and large multiplant firms than elsewhere. These larger firms do of course make extensive use of equipment; at the same time jobs tend to be highly specialized so that a somewhat smaller proportion of the work force than elsewhere operates equipment.

It is *not* true that the jobs of upper income people are least closely related to machinery.[7] Rather, very low income is most often associated with jobs which involve no equipment at all. Jobs which bring the worker into contact only with the equipment he operates also are held relatively frequently by the lower income and education groups. What stands out is that higher paid workers tend to report indirect use of machines much more often than those with lower incomes. Similar patterns are reflected in the educational breakdowns, although the relation of education to machine use is less pronounced than that of income to machine use. In all, so pervasive is the relation between job and machine that even among the college educated and among those with incomes over $10,000 over 40 percent reported that they *operate* machines at least some of the time. The frequency of high levels of income and education among machine users has essentially two reasons: (1) the increasing use of machines in offices of all kinds, professional service establishments, financial institutions, hospitals, research organizations, even libraries and schools, and (2) the growing importance of complex and sophisticated machinery. Fifty, or even twenty-five, years ago the image of the machine operator as typically a person of low income and educational attainment may have been correct; it is no longer correct today.

There is a relatively large proportion of young workers among those who operate equipment. Over 70 percent of those under age 25 operate equipment, but only 55 percent of those over age 55. Workers who have only indirect contact with machines are concentrated in the middle age range.

Negroes are more likely than white workers to have jobs which involve no use of equipment. They are found less frequently than white workers in jobs which involve indirect use of equipment. These differences are consistent with the lower educational level of Negroes and their concentration in unskilled and service jobs.

Who are the people who use automated equipment? How do they differ from those who use more traditional equipment? These questions can be explored on the basis of the Automation Scale presented in Table 2-3, Part B. The sharpest difference between people who use automated equipment and those who use more traditional equipment relates to education. The more

---

[7]It should be noted that all income figures in this study refer to the worker's personal earnings from his main job, *not* family income.

TABLE 2-5

AUTOMATION LEVEL OF EQUIPMENT USED IN DIFFERENT OCCUPATIONS

Occupation	Numerical, tape, computer or other logical control	Fixed mechanical control	Powered multi-system, manual main control	Manual control, operator powered or powered single-system	All	No. of cases
A. Automation level of equipment operated directly						
Professional and technical workers	14%	75%	3%	8%	100%	251
Managers, officials	23	67	4	6	100	99
Proprietors, self-employed businessmen	5	47	17	31	100	103
Clerical and sales workers	10	78	9	3	100	386
Craftsmen, foremen	1	38	22	39	100	328
Operatives	1	50	33	16	100	423
Service workers	*	57	4	39	100	137
Laborers - nonfarm	*	16	46	38	100	57
Farmers and farm workers	*	8	87	5	100	120
B. Automation level of equipment used indirectly						
Professional and technical workers	42%	53%	3%	2%	100%	127
Managers, officials	49	28	19	4	100	85
Proprietors, self-employed businessmen	*	60	18	22	100	49
Clerical and sales workers	46	44	8	2	100	138
Craftsmen, foremen	6	56	33	5	100	214
Operatives	3	70	21	6	100	128
Service workers and laborers	9	48	32	11	100	55

*Less than 0.5 percent.

mechanized or automated the equipment, the more years of schooling were reported by the people who work with that equipment either directly or indirectly (see Table 8-1). More than two-thirds of the people who work with computers or other logically controlled equipment are college trained. And even among those in the fixed mechanical control group (the category which includes the heavy, highly mechanized equipment) a third have at least some college education; two-thirds have completed high school. In jobs where equipment is manually controlled only about 13 percent of the workers are college-educated and at least 50 percent have not finished high school.

Associated with the educational differences among people at various levels of automation are less pronounced differences in income (see Table 5-1). The median income from the job of those who operate logically controlled equipment was $8,200 per year in 1966. For the remaining groups which use machines median annual job income was in the $5,300-5,800 range without showing any systematic relation to automation level. These findings raise larger issues regarding the relation of mechanization and automation to educational requirements and people's financial progress. In Chapters 5 and 8 these matters will be analyzed more thoroughly.

The relation of automation level to occupation is illustrated in Table 2-5. The two lowest and the two highest categories on the 6-point Automation Scale are combined in order to avoid groups with very small numbers of cases. Considering only jobs which require working with equipment,[8] the occupations which most frequently involve operation of logically controlled equipment are the white collar occupations—professional and technical, managerial and administrative, and clerical. These same groups also have an important indirect involvement with logically controlled equipment, much more often so than other occupations. When people in these three white collar categories make indirect use of equipment, over 40 percent of the time some logically controlled equipment is involved. At the other extreme, the occupations which most often employ manually-controlled, single-system machines are laborers, service workers, craftsmen, and foremen, and some of the self-employed (such as artisans).

Degree of automation is related to several other socio-economic characteristics (Table 2-6). These relations are illustrated only with reference to equipment operated by the respondent; parallel patterns are apparent when workers are classified by degree of automation of equipment with which they work indirectly. People in the middle age brackets work with highly automated equipment somewhat more frequently than the youngest and the oldest workers. Those who operate logically controlled equipment hardly ever

---

[8]The proportion in each occupation category not using equipment appears in Table 2-4, Part A.

TABLE 2-6

AUTOMATION LEVEL OF EQUIPMENT ASSOCIATED WITH SOME DEMOGRAPHIC CHARACTERISTICS

(Percentage distribution of those who use equipment)

	Automation level of equipment operated directly					
	Numerical, tape, computer or other logical control	Fixed mechanical control	Powered multi-system, manual main control	Manual control, operator powered or powered single-system	All	No. of cases
A. Age						
Under age 25	4%	64%	18%	14%	100%	280
25-34	9	53	20	18	100	437
35-44	7	53	22	18	100	455
45-54	4	52	23	21	100	425
Age 55 or older	1	54	25	20	100	317
B. Size of firm's employment[a]						
Multi-plant firm						
Less than 50 employees	5	64	19	12	100	306
50-499	6	55	21	18	100	367
500-1,999	10	58	15	17	100	174
2,000-4,999	10	63	9	15	100	103
5,000 or more	13	55	9	23	100	107
Single plant firm						
Less than 50 employees	3	47	28	22	100	544
50-499	5	59	18	18	100	154
500 or more	2	72	11	15	100	54
C. Change in size of firm						
More employees	8	61	15	16	100	846
Same	2	47	31	20	100	568
Fewer	5	47	29	19	100	204
D. Race						
White	6	56	21	17	100	1,709
Negro	2	44	28	26	100	167

[a]Number of employees in the respondent's plant or office.

are over 54 years old. As might be expected, automated equipment is found most frequently in the larger establishments of multiplant or multioffice firms, and interestingly, also in firms with expanding employment. White people tend to have use of automated equipment more often than Negroes, particularly when it comes to equipment which affects the job indirectly.

C. People's Attitudes Toward Machinery

It appears that American workers in all occupations like the machines they work with and feel that automation is a good thing for people like themselves. That machines may deprive people of a sense of achievement or individuality, of the opportunity to organize their own work or to enjoy the presence of co-workers seems to be an unfounded concern for the large majority of labor force members.

A number of questions were asked to explore attitudes toward machinery and automation in a rather broad sense. To start with, people who operate machines all or some of the time were asked:

"Some people tell us that the equipment they work with is like a friend that helps them to do something. Other people say that their equipment is more like a foe that they have to struggle with, or that it is even running their lives. How do you feel about the equipment on *your* job?"

Respondents had an opportunity to express themselves freely in their own words, and their views with all qualifications were fully recorded. The tabulation below shows that more than four people out of every ten who operate machines said flatly that the machine is a "friend." Another three in ten called it a "help" or pointed out how much easier the machine makes the job; a few others used the word "friend" but added some qualifications. In sum, three-fourths of the responses were clearly favorable.

Machine is like —

A friend	45%
A friend, qualified	28
Pro-con	14
A foe, qualified	3
A foe	3
Not ascertained	7
Total	100%

Most of the remaining people who operate machines pointed out that machines are a necessity; that they have good and bad features. Only about 6 percent expressed qualified or unqualified dislike of the machines they work with. These attitudes did not vary by the extent to which the worker's equipment was automated.

A rather different question was asked later in the interview:
"In general, would you say that automation is a good thing for people doing your kind of work, or does it cause problems, or doesn't it make any difference? Why do you say so?"

The answers are presented in Table 2-7 in relation to machinery used by the respondent. About 40 percent of people who work with machines and over 60 percent of others said that automation makes no difference as far as their job is concerned. A recurring comment by these people was: "My job *can't* be automated." This reply was obtained quite often from workers whose low—skilled jobs would seem to be quite plausible candidates for automation. Conceivably some suppressed uneasinesss about automation is present in this group.

Among those who did feel that automation matters, three out of every four expressed favorable attitudes, almost always unqualified. Some others gave pro-con answers. In the population as a whole a small minority, 11 percent, said that automation is a bad thing for people in their line of work. Even among those who operate machines all or most of the time, this figure was only 12 percent. It was 9 percent for those who don't work with machines and 10 percent for those whose contact with machines is only indirect. Workers who already use highly automated equipment are at least twice as likely as others to feel that automation is a good thing for people in their line of work; they are much less likely than others to say that automation makes no difference or is a bad thing (lower part of Table 2-7).

When asked to explain their favorable or unfavorable attitudes toward automation, people readily expressed reasons. Evidently, the word "automation" was interpreted quite loosely. Those who argued that automation is a good thing for people doing their kind of work most often spoke of the productivity of automated or mechanized equipment. They articulated this idea in many different ways, pointing out that automation raises the worker's efficiency, makes the work easier (physically or otherwise), increases output, lowers costs, improves or standardizes quality. Understandably, to many people it was inconceivable that their kind of work could be done without "automated" equipment.   About one worker in ten expressed favorable ideas such as these: Using automated equipment makes a worker feel important; it gives him pride in his job. Automated equipment enables people to do new things; it requires new skills; it makes the work more interesting. Automated equipment is safer. It makes for a cleaner and more agreeable work environment.

TABLE 2-7

ATTITUDES TOWARD AUTOMATION BY PATTERN OF EQUIPMENT USE
AND BY AUTOMATION LEVEL OF EQUIPMENT

	Automation is						No. of
	Good	Pro-con	Bad	Makes no difference	Not ascertained	All	cases
All people	34%	3%	11%	47%	5%	100%	2,662
Workers' relation to equipment							
Both operate and have indirect use	48	4	13	33	2	100	540
Operate only	32	3	11	49	5	100	1,127
Indirect use only	47	3	10	37	3	100	367
None	20	2	9	62	7	100	586
Automation level of operated equipment							
Numerical, tape, computer or other logical control	75	4	7	13	1	100	106
Fixed mechanical control	39	3	13	41	4	100	1,038
Powered multi-system, manual main control	33	2	10	50	5	100	409
Manual control, operator powered or powered single-system	25	3	12	55	5	100	348

A majority of the 11 percent who felt that automation is a bad thing for people in their line of work gave this one explanation: Automation replaces people by machines.

The impression that machines seldom detract from job satisfaction is reinforced when one compares more general attitudes toward the job among groups who do and those who do not work with machines. The question was:

"On the whole, do you feel that the work on your present job is drudgery, or is it all right, or do you enjoy your work?"

Three-fourths of all members of the labor force said that they enjoy their work and very few qualified this answer in any way. Only about 5 percent

TABLE 2-8

ATTITUDE TOWARD JOB BY PATTERN OF EQUIPMENT USE
AND BY AUTOMATION LEVEL OF EQUIPMENT

	Job is				
	Enjoyable	Pro-con	Drudgery	Not ascertained	All
All people	77%	17%	5%	1%	100%
Workers' relation to equipment					
Both operate and have indirect use	81	14	4	1	100
Operate only	76	18	5	1	100
Indirect use only	78	15	6	1	100
None	76	17	6	1	100
Automation level of operated equipment					
Numerical, tape, computer or other logical control	91	8	1	*	100
Fixed mechanical control	78	16	4	2	100
Powered multi-system, manual main control	75	20	4	1	100
Manual control, operator powered or powered single-system	76	19	5	*	100

said that their work is drudgery or close to drudgery. This finding is consistent with survey answers obtained in many fields of human activity, which consistently show that a substantial majority of people tend to express satisfaction with their possessions and major activities (whether it's their marriage, their house, their income, or their refrigerator). What interests us here are any differences in expressed job satisfaction between people who do and those who do not work with machines. The upper part of Table 2-8 indicates no significant differences between groups classified by machinery use. The lower part of the table does, however, suggest that workers with automated equipment tend to enjoy their work more often still than those with more traditional equipment. Hardly anyone working directly with the most highly automated equipment feels that his work is drudgery.

It was recognized that, in order to uncover job dissatisfaction due to machinery use, the interview might have to be more suggestive. Two potentially negative consequences of machinery use are the lessened opportunity to talk to fellow workers and to organize one's own work. Questions were therefore asked of all wage and salary earners regarding both these matters. About 70 percent of people replied flatly that, where they work, they can talk any time (Table 2-9). However, the percentage who are unable to talk as much as they want does show some weak relation to machinery use. It ranges from 26 percent for those who operate machines to 13 percent for those who don't work with machines; and it is 7 percent for those who use equipment only indirectly. The most frequent reasons given for not being able to talk (or for sometimes not being able to talk) is that no fellow worker is close enough or that the job requires too much concentration. A few workers explained that there is too much noise.

With regard to work organization, about 50 percent of the labor force feel that how their work is organized depends mostly on themselves (Table 2-9). For people who operate machines this figure is 44 percent, for those who work with equipment they do not operate it is 55 percent, and for those who do not work with equipment at all 52 percent. Another 18 percent said that work organization is *partly* determined by themselves or by a group of people working together. Only 16 percent of wage and salary earners felt that work organization is determined wholly or in part by equipment or by a production line or a machine-paced work flow. Among people who operate machines this last figure is 19 percent; among people who have no contact at all with machines it is only 12 percent.

Another common charge against machines is that they underutilize human capabilities and skills and reduce job satisfaction on that account. People who operate machines were asked:

"Thinking now about *just the equipment or machinery you use* on your job, how long would it take someone with the same *general* education and training as you to learn how to work properly with this equipment?"

More than 20 percent of people who operate machines mentioned periods shorter than 8 hours and about one-half mentioned periods shorter than half a month. At the other extreme, a small minority, about one-sixth of the group, spoke of periods in excess of 8 months. The accuracy of these judgments is open to question. Regardless of accuracy, the training required to work with a piece of machinery is different and is of a lower order than the capabilities demanded by the job as a whole. We shall see in Chapter 6 that on balance, workers who have experienced an advance in machine technology see their total job as having become more demanding, despite the small amount of training needed to work with the new machines.

A question regarding outmoded skills was asked of all members of the labor force:

TABLE 2-9

WORKERS' PERCEPTIONS OF THEIR OPPORTUNITY TO TALK AND
TO ORGANIZE THEIR WORK, BY PATTERN OF EQUIPMENT USE

Do workers have a chance to talk?	All[a]	Both operate and have indirect use	Operate only	Indirect use only	None
		Workers' relation to equipment			
Yes, can talk any time	69%	68%	67%	78%	73%
Talking is part of the job	8	6	5	12	12
Sometimes cannot talk	9	11	11	1	8
No, cannot talk	11	12	15	6	5
Not ascertained	3	3	2	3	2
Total	100%	100%	100%	100%	100%
How is work organized?[b]					
By equipment, production line, work flow	16%	15%	19%	14%	12%
By worker himself	48	46	44	55	52
Group of people	18	19	19	17	18
By boss, supervisor	10	11	11	5	10
Several things, other	5	6	5	6	6
Not ascertained	3	3	2	3	2
Total	100%	100%	100%	100%	100%

[a]Includes those for whom kinds of equipment were not ascertained.

[b]These questions were not asked of the self-employed.

The questions asked were: "Do you have much chance to talk to other people during the time you are working, or does your location or the kind of work you do keep you from talking with others while you work?" "In some jobs, how your work is organized depends mostly on you, yourself. In others, it's pretty much determined by the equipment you work with, a production line, or by a group of people that have to work together. How is it with your job?"

TABLE 2-10

PERCEIVED USE OF SKILLS
BY WORKERS, IN RELATION TO EQUIPMENT USE

| | Workers' relation to equipment | | | | |
	All[a]	Both operate and have indirect use	Operate only	Indirect use only	None
Perceived use of skills					
Have unused skills	17%	21%	19%	11%	14%
Don't have unused skills	82	79	80	89	85
Not ascertained	1	*	1	*	2
Total	100%	100%	100%	100%	100%

[a]Includes those for whom machine use was not ascertained.

"Through your previous experience and training, have you built up some skills that you would like to be using, but can't on your present job?"

Table 2-10 shows that people who operate machines are more likely to feel that they have unused skills than others, although the differences are not very large, and the frequency of reports of unused skills is not high even among people who operate machiney.

In all, by directing people's attention to potential negative consequences of machinery use—not being able to talk, to organize one's own work, or to use all one's skills—one can find small groups in the labor force which do seem to feel such adverse effects. The chance of job dissatisfaction appears to be marginally increased by work involving machine operation, as against work involving only indirect use of machines or no machine use at all. Still, what needs to be emphasized is that the large majority of American workers like to use machines; whatever the unfavorable aspects of machine use may be, in most cases they seem to be of little salience when people speak of or evaluate their jobs.

In chapter 6 we shall analyze changes in attitudes and perceptions which accompany *changes* in machine technology, to explore further how machines add to or reduce job satisfaction.

                    *   *   *

In this chapter we have examined some of the attitudes and economic experiences associated with machinery *use.* We found that members of the U.S. labor force overwhelmingly like the machines they work with and enjoy jobs which involve the use of equipment, either directly or indirectly. There are only a few traces of job dissatisfaction or frustration associated with machinery use. The major focus of this study is on the impact of machinery *change.* In view of the findings of this chapter, it would not be surprising if workers for the most part reacted favorably to advances in machine technology on their job. This is the subject to which we now turn.

Chapter 3

CHANGE IN MACHINE TECHNOLOGY

In this chapter we shall turn to the central problem of this study—*change* in machine technology over a 5-year period and its impact on the labor force. Section A will discuss some measurement problems. In Section B we shall indentify the groups which reported that they experienced changes in machine technology on their jobs and estimate the frequency of such occurrences. The kinds of machinery changes which took place will be described briefly in Section C. Finally, Section D will present an analysis of the personal and job characteristics of people who were affected by changes in machine technology during the past 5 years.

A. The Measurement of Change in Machine Technology

Change in machine technology can be defined in different ways, and definitions and methods of measurement will obviously affect the frequency with which such change is observed.[1] First of all, there is no doubt that we are interested in cases where a worker reports a change in a machine he operates. Accordingly, all members of the labor force were asked:

"During the last 5 years was there any *change* at all in any equipment or power tools which you yourself have operated or helped to run on your job? Was there any change in equipment which is directly connected to any equipment which you have operated or helped to run?"

_____

[1]See also Section A on "Defining Change in Machine Technology," Chapter 1 above.

Secondly, a person's job content may be greatly altered by changes in equipment which he himself does not operate. We saw in Chapter 2 that a good many people reported that their work is significantly affected by equipment which they do not operate. Thus the further question was asked:

> "Now, thinking again about equipment that you *don't* operate but which is important for your work—I mean equipment in your company or even in other companies—during the last 5 years was there any *change* at all in equipment having an effect on how you do your work?"

Self-employed people who introduced changes in the equipment with which their employees work also fall into the category of workers who experienced changes in equipment to which they are related only indirectly; they were asked a special question referring to this possibility. Accordingly our definition of change in machine technology has two categories (often shown separately in the tables)—changes in the worker's own equipment and changes in other equipment that affects his work indirectly.

A second measurement issue pertains to the *importance* of changes reported by survey respondents in reply to this sequence of questions about possible machinery change. This problem was approached in two steps. The initial questions (quoted above) were worded so as to elicit reports about all changes, large or small. The intent was to cast a wide net and to minimize omissions of experiences which might conceivably be relevant. No doubt, some rather trivial changes were mentioned in reply to the initial questions. The second step was designed to screen out inconsequential changes. It consisted of a probe which followed each mention of a change in machinery affecting the respondent's work either directly or indirectly:

> "Would you say that what *you* had to do on your job was changed quite a bit by the new equipment, was it changed somewhat, or did you keep on doing nearly the same thing in nearly the same way?"

Regarding the answers to this question, it should be noted that we are *not* interested in the extent of the technolgical change or the change in the equipment, but in the extent of change in the respondent's work. These dimensions of change may be correlated in some cases, but in others they will differ. No doubt people's answers about the degree of change in their work have a subjective element. Occasionally, a time-and-motion expert might make a different judgment. The worker's supervisor might or might not agree with the worker's evaluation of the importance of the change. The judge of magnitude of change in this study is the worker himself. The first table in this chapter will present data which include all workers who reported *any* changes in machine technology during the past 5 years. Thereafter the analysis of the impact of changes in machine technology will exclude workers who experienced only changes which left them doing "nearly the same thing" on their job. In other words, we are studying only changes in machine technology which affected people's work to some significant degree.

At the very outset of this study—during the process of questionnaire design and pretesting—it became evident that workers experienced changes in the machinery they operate in a number of different contexts, which could not be lumped together. For example, many people changed jobs during the 5-year period under study. On the new job they may have encountered the same or different equipment from that which they used on the former job. The job change may or may not have been precipitated by automation of their former jobs. People who were transferred from one department or one plant to another are similar to job changers. Again, a change in equipment may be the cause or the result of the transfer (or both or neither). The self-employed differ from wage and salary earners in that for them the change in equipment was discretionary, rather than being decided upon by a superior. In addition, the change-over to the new equipment was carried out according to their own plans and preferences. It was decided accordingly to distinguish between four different groups of workers: (1) Wage and salary earners who worked for the same employer during the 5 years prior to the interview and were not transferred between sections, departments, or plants; (2) wage and salary earners who stayed with the same employer during the previous 5 years but experienced one or more transfers; (3) wage and salary earners who had one or more job changes during the previous 5 years; (4) the self-employed. Separate questionnaire forms were used for these four groups. Whenever possible the questions in the four schedules were identical or very similar. However, some questions were inapplicable to one group or another, some were relevant only to one group, and some had to be modified to fit different circumstances. In the analysis in this chapter the four groups are sometimes viewed separately, and sometimes jointly in order to obtain groups of sufficient size for detailed analysis. In particular, the questionnaires for the people with job changes and transfers are sufficiently alike to justify combining these two groups for many purposes.

Occasionally, a worker went through several changes in machine technology during the past 5 years. In such cases he was asked to think about "the change which had the greatest effect on you and your work" and to answer with respect to this important change.

With these definitional issues out of the way, we may now turn to the central question of this study: What proportion of the labor force experienced changes in machine technology during the past 5 years?

B. *Frequency of Experiences With Change*
   in Machine Technology

Table 3-1 presents two distributions. Both show the labor force classified according to their reported experiences with *change* in machine technology. In the first column all changes in machinery are tabulated, however inconsequential they may have been for the respondent's work. In the second column only those equipment changes are taken into account which modified the respondent's job content to some appreciable degree. Therefore the proportion of the labor force who were affected by changes in machinery appears considerably larger on the basis of column (1) than of column (2). According to Table 3-1, 33 percent of all members of the labor force underwent some change in the machinery they work with during the past 5 years, while 22 percent experienced a change which they said altered their work to some appreciable extent. This latter group, 22 percent of the labor force, experienced change in machine technology in a variety of contexts, which we shall now examine.

About 51 percent of the labor force consists of wage and salary earners who worked for only one employer in the previous 5 years *and* remained in the same department or section. Not surprisingly, this group does not show as high a frequency of equipment change as people with job changes and transfers. If those employees whose work was not modified significantly by the equipment change are put into the "no change" category (column 2 of Table 3-1), the group includes 44 percent of the labor force with no equipment change, 4 percent whose work was changed in the past 5 years by innovations in equipment which they operate, and another 3 percent whose job content was changed only by innovations in equipment with which they work indirectly.[2] In other words, equipment change was experienced in a static employment context by 7 percent of the labor force, or by 14 percent of wage and salary earners with no job change or transfer.

Similar to the people with no employment change are the self-employed, who make up 12 percent of the labor force. About one in six of the self-employed—2 percent of the labor force—made some change in machine technology which affected their own work. Occasionally their work was altered by machinery changes introduced by their suppliers or customers. Almost all the machinery changes which affected the work of the self-employed were changes in machinery which they themselves operate at least part of the time.

---

[2]In a very small number of cases the changed machinery is actually located in another company, for instance where a company uses the computer services of an outside firm or where a doctor sends the results of medical tests to an independent laboratory for analysis.

TABLE 3-1

DISTRIBUTION OF THE LABOR FORCE BY CHANGES IN
MACHINERY AND JOBS OVER THE PAST FIVE YEARS

	All changes	Changes which had a significant effect on job	Equipment changes which had a significant effect on job
Same employer			
Change in equipment both operated directly and used indirectly	2%	1% 1%	
Change in equipment operated directly	4	3 3	
Change in equipment used indirectly	4	3 3	
No change in equipment	41	44	
Self-employed[a]			
Change in equipment both operated directly and used indirectly[b]	*	*	
Change in equipment operated directly	2	2 2	
Change in equipment used indirectly[b]	1	*	
No change in equipment	9	10	
Change in employer or transfer			
Change in employer or transfer because of equipment change	1	1 1	
Change in equipment	19	12 12	
Different employer		10	
Transfer		2	
No change in equipment	17	24	
Different employer		18	
Transfer		6	
Total	100%	100%	22%

*Less than 0.5 percent.

[a] In this study no persons were found who were now self-employed and who had left a previous job because of an equipment change.

[b] Equipment used indirectly may refer to equipment used by employees but not by the self-employed owner, or to equipment used indirectly by the self-employed owner.

Table 3-1 further shows that somewhat more than half the workers who experienced a change in machine technology did so in connection with a transfer within the same company or with a change of employers. Altogether about 37 percent of the labor force reported that they had changed jobs or were transferred within the past 5 years (28 percent were job changers and 9 percent transfers). This large group includes a small minority, about 1.2 percent of the labor force, who reported that they changed jobs or were transferred *because of* a change in machine technology. Sometimes a technical change in the place where the worker was formerly employed led to the job change or transfer; sometimes new machinery in another company or department led to the change. A reading of the interviews indicates that most of these job changes and transfers were involuntary on the worker's part.

The various groups with machine change which we have examined so far add up to 10 percent of the labor force. The remaining 12 percent who experienced a change in machine technology did so in connection with a job change which, according to their own reports, was not caused by technological change. It might appear that job changes and transfers are a major mechanism by which jobs are filled that have been created or given different work content by new machinery. This inference would not be entirely correct. Some job shifting involving change to a different kind of equipment would occur even in a technologically static economy. We shall see later that there is no evidence that in the aggregate job changers work with more modern, more mechanized, or more automated equipment after the job change than before.

On the other hand, one may well question the subdivision of job changers into 1.2 percent who did and 36 percent who did not leave their former job because of technological change. This subdivision is based on people's own explanations of the reasons for their job shift. These reasons were elicited by an open-ended inquiry about reasons for the job shift, followed by a specific probe: "During the last 5 years, did you leave (any of) your former employer(s) because he changed to new machinery or because of automation on your former job—did that have anything to do with your leaving?" Most job changers referred to insufficient pay, short hours, lack of opportunity to advance, and unsteady work on their former job. Without the job changer knowing it, these conditions may be due to competition from technologically more advanced firms, so that some of these job changes would not be wholly unrelated to changes in machine technology. This supposition is reinforced by the finding that about one-fourth of job changers reported (in reply to a direct question) that their employer was cutting back production or reducing the work force at the time they left their previous job. Not surprisingly, the further probe—"Why was it that the work force or production was being reduced where you worked?"—did not always bring forth knowledgeable answers. Many people explained that the cutbacks were seasonal or cyclical, and the lay-off season is a good time to look for a more steady job.

A reading of interviews disclosed that in a good many instances the employers which cut back or went bankrupt ran small groceries, service establishments, and construction firms, which are known to have high mortality rates. There were a substantial number of reports of plant relocations, and some instances of cancellation of government contracts. In these kinds of cases change in machine technology has nothing or very little to do with the cutbacks in work. In a smaller number of cases the job changer merely stated that his former employer lost business or contracts; that he was no longer competitive; or that demand for his product fell off. In some of these latter cases technological change by competitors may have been at the root of the former employer's problems. Still, it would seem that in a large majority of cases equipment change is a result, but not a cause, of job change.

If we disregard the 12 percent of machine changes occasioned by job changes that were unrelated to technological advance, the proportion of the labor force whose work was clearly affected by technological change over a 5-year period may appear small—10 percent plus an unknown, but small, proportion of the job changers. Some further considerations *do* lend credence to this estimate. Rather dramatic changes in *equipment* often have a major effect on the *work* of only a small proportion of the employees in a company. Clerical workers in particular often perform a variety of tasks. Some of these tasks are transferred to new machines, but many employees will continue to perform the remaining accustomed duties which may be growing in volume. Einar Hardin found this to be the case in his study of an insurance company that installed an IBM 650 electronic data processing machine in 1957. The computer fully automated the checking of claims and the processing of insurance policies. Yet, 60 percent of home office employees reported no change at all in their job as a result of the computer installation and another 11 percent reported only a "slight" change.[3] It is the essence of highly mechanized or automated equipment that it requires few operators. It creates "new" kinds of jobs, but only in small numbers. When technological advance occurs in a setting of vigorous economic expansion, as it did in 1962-67, the increased output of goods and services implies a growing need for people who will perform the many conventional tasks which have not been mechanized or automated, i.e. tasks which are not "new."

Our estimates also are affected by entries into and retirements from the labor force. We saw in Chapter 2 (Table 2-6) that young people are somewhat more likely than older people to operate machines under fixed mechanical or logical control. We may assume then that young people who enter the labor force have some tendency to gravitate toward jobs which require use of technologically new equipment, while those who retire tend to leave jobs which involve more traditional equipment. Office automation commonly affects young women who, for reasons of marriage and family, may stop working after a few years. Their experiences then have been omitted from our survey, which includes only present members of the labor force. The concern of this survey is the millions of men and women who must adapt to the changing machine technology in the course of their working lives.

---

[3]Einar Hardin, "The Reactions of Employees to Office Automation," U.S. Department of Labor, Bureau of Labor Statistics, Bulletin No. 1287, *Impact of Automation*, 1960, pp. 101-108.

Although the incidence of machine changes which alter people's work significantly may appear low when stated as a percentage of the labor force, in absolute terms large numbers of workers are involved. It may be estimated from Table 3-1 (column 2) that nearly 15 million workers (22 percent of a fulltime civilian labor force of 67 million) experienced an equipment change which altered their job content during the five years under study. Of these about 7 million workers were clearly affected by technological advance. This figure includes nearly a million who reported that they changed jobs *because of* a change in machine technology. It includes a further 2.7 million members of the labor force who experienced a change in the equipment they operate while remaining on the same job. It covers nearly 2 million, also remaining on the same job, who underwent a change in equipment which related indirectly to their work. Finally, it encompasses about 1.3 million self-employed whose work was affected by changes in machine technology. The remaining 8 million workers changed jobs or were transferred, presumably for reasons unrelated to technological advance, *and* had to work with different equipment on the new job. Perhaps technological change was indirectly responsible for another million or so of these job shifts.

It appeared in Chapter 2 that many people's work is affected only indirectly by machinery; and there are others who report that their work is entirely unrelated to equipment. It is interesting therefore to examine the frequency of equipment change in relation to equipment use. In doing this we shall confine ourselves to equipment changes which altered people's work (those shown in columns 2 and 3 of Table 3-1). While 22 percent of the entire labor force underwent some change in equipment during the past 5 years, of the people who operate equipment "almost constantly" 28 percent experienced a change in the equipment they operate and another 2 percent reported that their job content was altered by changes in equipment they do not operate. The corresponding frequencies for people who operate equipment only "some of the time" were 25 and 3 percent; and for those who operate equipment "very little" they were 15 and 3 percent. Evidently, the more continuously a worker operates equipment, the more likely he is to feel that his work was altered significantly by a machinery change. It is also likely that people who operate machinery all the time use a large number of pieces of equipment; hence the chance for change is enhanced. A very small fraction of labor force members who now operate no equipment also reported a change in machine technology which affected their work. Apparently these people switched at some time during the past 5 years from work which entailed some machine operation to work which entailed none, possibly but not necessarily as a result of changes in machine technology. Among the group which has only indirect contact with machinery 23 percent reported a change in machinery which altered their work.

In order to avoid misunderstanding, it is worth repeating here that by no means all the machine changes measured imply "automation" in the strict sense of the word. Some do; others may represent increased mechanization of equipment which remains essentially under manual control, or shifts from manual to fixed mechanical control. We may assume that in cases where no job change is involved, all machine changes represent technological advance (a firm hardly ever shifts to less modern equipment). Where job changes are the occasion for machine change, the employee may change to equipment on either a higher or lower level on the Automation Scale; there is bound to be a mixture of movements in both directions.

C. Kinds of Technological Change

We shall now look briefly at the kind of machinery change which was reported. Not surprisingly, people could describe work changes much better than machinery changes. Only those data pertaining to machine change which appear reliable and meaningful are worth discussing. People *were* able to identify or name the pieces of equipment that were changed or newly in-troduced.   Table 3-2 shows—for people who had no job change or transfer— a distribution of the equipment which they used at the time of the survey and also a distribution of the equipment which was changed or newly added in the 5 years prior to the survey. Equipment which the respondent operates is tabulated separately from equipment which affects his work indirectly.[4] In contrast to most other tables in this report, Tables 3-2 and 3-4 present dis-tributions of equipment rather than of people. That is, where several pieces of equipment were used by one worker or were changed, each piece of equip-ment is a separate case.

The table shows what one might expect: more changes were reported for the more modern and complex types of equipment than for the more traditional and simpler machines. The high frequency of changes in computers and electronic office equipment stands out in Table 3-2. Computers and large-scale electronic office equipment account for over a quarter of the changes in equipment operated. Another quarter of the changes pertain to power tools and other small-scale production equipment. However, while changes in com-puters and electronic office equipment were frequent relative to the use of such equipment, this is not true for power tools and small production equip-ment.   Changes in this latter category were frequent only because use of such equipment is so widespread (30 percent of the equipment people operate

---

[4]Only equipment changes which affected people's work appreciably are included in Table 3-2.

TABLE 3-2

COMPARISON OF ALL MACHINERY USED AND MACHINERY THAT WAS CHANGED[a]

(Percentage distribution of machinery used by people
who had the same job for the past 5 years)

Type of machinery	Equipment operated directly		Equipment used indirectly	
	All	Machinery that was changed	All	Machinery that was changed
Transportation vehicles--all types, passenger and freight	7%	3%	7%	2%
Small-scale office equipment	25	13	10	3
Tools, small-scale production, household, institutional equipment	30	25	13	7
Mobile equipment--construction, farm, freight	5	5	6	5
Computers, electronic equipment	7	29	30	53
Production equipment--large scale, non-mobile	16	19	30	25
Professional, specialized equipment	4	4	3	2
Miscellaneous equipment	6	2	1	3
Total	100%	100%	100%	100%
Number of cases	(1,476)	(122)	(542)	(106)

[a]Columns 1 and 3 of this table differ from the corresponding columns in
Table 2-3 in that only people who had the same job for the past 5 years
are included here.

is of this nature). Traditional small-scale office equipment, though very
commonly used, accounted for a relatively small proportion of the reported
changes. The frequency of changes in large-scale production equipment cor-
responds roughly to the proportion of this equipment in the total of reported
equipment.

The same pattern is evident when one looks at changes in equipment
which affects people's work only indirectly (columns 3 and 4 of Table 3-2). If
anything, the impact of computers and electronic office equipment on
people's work appears even more striking in this context. Over one-half of
these machinery changes involved computers or other electronic equipment.

A further question concerns the extent to which reported machinery
changes were labor-saving. People who experienced significant changes in the
equipment they operate were asked:

"Were more or fewer people needed to put out the *same amount of work* per day, or was there no change in this respect?" (If the answer mentioned change) "Was this a large change or a small change?"

Table 3-3 (column 1) shows that in about 40 percent of cases, the respondent reported that fewer people were needed to do the same amount of work. And, more often than not, the employee described the reduction in the amount of labor needed as a "large" one. The same question also was asked of people who experienced a change in the equipment with which they have only indirect contact. In this case somewhat over one-half reported that fewer people were needed (Table 3-3, column 2). That technological change is not always labor-saving should not be surprising. Often improved machines serve the purpose of turning out a new or better product. On production lines the work is often made easier and more agreeable, without involving labor replacement. Indeed, in a few cases respondents reported that the new equipment requires more workers. Those who used a computer for research purposes sometimes explained that it enables them to obtain more information (in part however by being labor-saving). The same point was made by a department head in a credit rating service and by medical people who can make additional tests to diagnose their patients' complaints. Teachers who have come to use movie projectors, reading machines, language laboratory equipment and the like also offer essentially improved services. We must of course recognize that the expression *the same amount of work per day* has no precise meaning where the new machine is associated with an improved product or service. Thus the estimate obtained from respondents, that in a little less than half the cases important equipment changes were labor-saving, may be on the low side.

We may now look at the kinds of equipment changes experienced by workers who had job changes or transfers during the past 5 years. Table 3-4 shows the equipment used by these people before and after the employment shift. When Table 3-4 is compared with Table 3-2 only minor differences can be discerned in the equipment worked with by people who stayed on the same job during the past 5 years and those who changed jobs or were transferred. In other words, it is *not* true that job changes or transfers are more frequently associated with modern than with traditional equipment, or vice versa. Perhaps some people leave the more automated jobs because they find the work uncongenial; others may leave jobs with more traditional equipment seeking greater challenge and opportunity for advancement. In most cases type of equipment used has little or nothing to do with the decision to leave a job. Most interestingly, Table 3-4 indicates that overall the distribution of equipment used before the employment shift is quite similar to that used after the shift. The majority of people remain in the same general occupation when they change jobs or are transferred. The changes in equipment experienced by

TABLE 3-3

LABOR REQUIREMENTS OF CHANGED EQUIPMENT[a]

New equipment requires:	Equipment operated directly	Equipment used indirectly
Less labor per unit of output	40%	51%
Large change	23	24
Small change	17	27
No change	56	42
More labor per unit of output	4	7
Total	100%	100%
Number of cases	(262)	(171)

[a]Includes people who had the same employer and a change in equipment.

TABLE 3-4

COMPARISON OF MACHINERY USED BEFORE AND AFTER JOB CHANGES AND TRANSFERS

(Percentage distribution of machinery used by people who changed jobs
during the past 5 years and who experienced a change in equipment)

Type of machinery	Equipment operated directly		Equipment used indirectly	
	Before job change	After job change	Before job change	After job change
Transportation vehicles--all types, passenger and freight	9%	8%	8%	11%
Small-scale office equipment	19	23	11	9
Tools, small-scale production, household, institutional equipment	24	25	11	11
Mobile equipment--construction, farm, freight	9	6	6	6
Computers, electronic equipment	7	7	25	26
Production equipment--large scale, non-mobile	19	20	30	34
Professional, specialized equipment	4	5	5	2
Miscellaneous equipment	9	6	4	1
Total	100%	100%	100%	100%
Number of cases	(362)	(558)	(178)	(189)

job changers and transferees may not be so extensive that they imply a major reclassification of the equipment, although all the reported changes presumably are ones which affected people's work appreciably. At the same time there are bound to be a good many offsetting changes: one worker moves from traditional to electronic office equipment, and another makes an opposite move. Both may have made an adjustment which to them implies more congenial work.[5]

Table 3-5 classifies equipment used before and after job shifts and transfers by automation level. The table confirms the impression conveyed by Table 3-4: In the aggregate, job changes and transfers do not involve a significant shift toward more mechanized or more automated equipment; there may however be some upward shift *within* our rather broad automation categories. The earlier supposition that much of the machine change that is incidental to job changes and transfers might be observed even in a technologically static economy is reinforced by these data. Moves to more sophisticated equipment are balanced by moves to less sophisticated equipment among workers with transfers and job changes.

D. Who Experienced Changes in Machine Technology?

Workers who reported that they experienced changes in machine technology which affected their work differ from others both with regard to personal characteristics and with regard to the kind of companies they work for. On the whole, those who were affected by technical change during the past 5 years are in a stronger position in the labor market than those who worked with the same technology throughout (or have no contact at all with machines); but the differences which emerge are not very pronounced or consistent.

One characteristic of those who experienced technological change is their youth. Looking first at the group who held the same job during the past 5 years, 55 percent of those who were affected by technological change were in the 25 to 44 age bracket, but only 39 percent of those who reported no change were that young. We may disregard the group under age 24 here, since many of them have not yet been in the labor force for 5 years (hence the chance of a change is reduced). Perhaps when new machines are introduced, the employer selects younger workers to do the new work that is required. Perhaps young people select jobs which are technologically more dynamic. It is well known that young people change jobs much more often than older people. And Table 3-6 indicates that, when they do change, those under age

---

[5]No breakdown of the kinds of machinery changes experienced by the self-employed is presented here because of the rather small number of such cases.

TABLE 3-5

AUTOMATION LEVEL OF EQUIPMENT OPERATED BY PEOPLE WHO CHANGED JOBS

Automation level of equipment operated directly	Current equipment	Equipment on past job
Numerical, tape, computer or other logical control	7%	8%
Fixed mechanical control	57	59
Powered multi-system, manual main control	18	18
Manual control, operator powered or powered single-system	18	15
Total	100%	100%
Number of cases	(745)	(672)

TABLE 3-6

DISTRIBUTION OF WORKERS BY AGE AND JOB AND MACHINE CHANGE
DURING THE PAST 5 YEARS

Age	Same job for last 5 years and		Different job and		Self employed and		Total	No. of cases
	Machine change	No machine change	Machine change	No machine change	Machine change	No machine change		
Under age 25	3%	37%	27%	31%	*	2%	100%	347
25-29	5	29	28	30	3	5	100	278
30-34	8	30	18	35	1	8	100	294
35-44	9	45	12	23	2	9	100	643
45-54	6	52	6	20	3	13	100	610
55-59	6	52	6	17	2	17	100	235
60-64	7	61	2	13	1	16	100	159
65 or older	2	44	7	14	4	29	100	86

*Less than 0.5 percent.

29 are much more likely than older workers to accept a position which involves working with a different kind of equipment. After age 44 the probability that a job change involves an equipment change declines further. In contrast to job changers (who are young), the self-employed tend to be somewhat older than the rest of the labor force. In this group again the younger members more often reported technological change than the older ones.

People who experience changes in machine technology on their job also tend to have a higher level of formal education than other members of the labor force. Starting again with the group who held the same job for the past 5 years, 56 percent of those whose work has changed because of a change in equipment have a high school degree *and* either college or vocational training; the corresponding figure for those who reported no equipment change is 35 percent. However, among job changers, those who work with different equipment have the same educational distribution as those who work with the same equipment on the new job. Among the self-employed also, education is unrelated to technological change (Table 3-7).

Although women members of the labor force do work with equipment as often as men, they apparently work with equipment which changes a little more slowly than the equipment men use. A somewhat smaller proportion of female than of male members of the labor force reported that their work was affected by changes in machinery.

TABLE 3-7

DISTRIBUTION OF WORKERS BY EDUCATION AND JOB AND MACHINE CHANGE
DURING THE PAST 5 YEARS

Education	Same job for last 5 years and		Different job and		Self employed and		Total	No. of cases
	Machine change	No machine change	Machine change	No machine change	Machine change	No machine change		
0-7 grades	5%	53%	10%	19%	1%	12%	100%	212
8-11 grades[a]	4	50	14	20	2	10	100	763
High school degree[a]	7	41	14	25	2	11	100	854
Some college[b]	9	37	18	28	1	7	100	407
B.A. degree or higher degree	7	43	10	27	2	11	100	383

[a]Includes those who may have had some vocational training of any length in addition to the indicated high school training.
[b]One to 3 years of college, including all degrees below the B.A. level.

Changes in machine technology were reported by workers in all occupations and industries. Among wage and salary earners who had the same job during the past 5 years, white collar workers and skilled craftsmen show a somewhat higher incidence of technological change than unskilled and service workers (Table 3-8). Among job changers, on the other hand, operatives and laborers—the least skilled groups—had a particular propensity to take jobs which required them to work with a different kind of machine. These people, apparently because of their lack of skills, are more inclined than others to shift between occupations and industries when they change jobs. Frequent change in equipment is entailed in the kinds of job shifts made. Among the self-employed, farmers stand out as a group which reported changes in machine technology with particular frequency. The rapid rate of technological advance in farming is well known.   Among industries, manufacturing ranks first as regards frequency of equipment change; however much of this change occurred in connection with job changes, which are particularly numerous in manufacturing.

Negroes have a rather high degree of job mobility and experienced equipment changes in this connection as often as white workers. Those Negroes who remained on the same job throughout the past 5 years reported relatively little technological change. This finding is consistent with the low incidence of technological change among unskilled and service workers.

There is some tendency for workers employed by large employers, particularly those in multiplant firms, to experience more change in machine technology than those who work in small establishments (Table 3-9). This is the case in part because manufacturing has somewhat more technological change than service establishments, which are usually smaller (but may be part of a chain). More important, large firms are probably more progressive than smaller ones and in a better financial position to introduce new machinery when improved models become available.

Of particular interest is the finding that workers in plants where employment has been expanding (according to the worker's report) are more likely to have experienced changes in machine technology than workers in plants with stable employment (Table 3-9). The interviews suggest that firms which have an expanding market and are hard-pressed to find more labor are inclined to introduce new machinery or to modernize equipment. It should also be true, though this is probably less important, that technological advance gives a firm a competitive advantage. And, as we have seen, a good many technological changes are not labor-saving. Workers in firms with declining employment also experienced technological change with relative frequency. However, four times as many people reported increasing employment in their firm as reported decreasing employment. It appears then that in a majority of cases reductions in employment due to automation of equipment are more than offset over a

TABLE 3-8

DISTRIBUTION OF WORKERS BY OCCUPATION, INDUSTRY, AND JOB AND MACHINE CHANGE

Occupation	Same job for last 5 years and		Different job and		Self-employed and		Total	No. of cases
	Machine change	No machine change	Machine change	No machine change	Machine change	No machine change		
Professional and technical workers	9%	45%	10%	29%	1%	6%	100%	398
Managers, officials	9	51	9	31	*	*	100	185
Proprietors, self-employed businessmen	*	*	*	*	14	86	100	148
Clerical and sales workers	8	49	14	27	*	2	100	531
Craftsmen, foremen	10	44	17	24	1	4	100	421
Operatives	5	48	21	23	*	3	100	508
Service workers	1	53	10	30	2	4	100	222
Laborers--nonfarm	3	41	27	25	*	4	100	92
Farmers and farm workers	1	27	3	10	16	43	100	137
Industry								
Agriculture, forestry, fisheries, mining	4	30	4	11	13	38	100	166
Construction	5	35	17	24	2	17	100	180
Manufacturing	7	45	20	25	1	2	100	805
Transportation, communication, utilities	10	44	15	22	1	8	100	193
Trade--wholesale, retail	4	39	9	27	4	17	100	402
Finance, business services	9	39	11	28	1	12	100	196
Repair, personal, and entertainment services	5	45	11	17	4	18	100	140
Health, education and welfare services	6	55	8	26	1	4	100	386
Government	10	50	15	25	*	*	100	174

*Less than 0.5 percent.

TABLE 3-9

DISTRIBUTION OF WORKERS BY SIZE OF FIRM AND CHANGE IN FIRM'S EMPLOYMENT AND JOB AND MACHINE CHANGE

Size of firm's employment	Same job for last 5 years and		Different job and		Self-employed and		Total	No. of cases
	Machine change	No machine change	Machine change	No machine change	Machine change	No machine change		
Multiplant firm[a]								
Under 50 employees	9%	46%	13%	32%	*	*	100%	421
50–499	7	47	19	27	*	*	100	523
500–1,999	10	51	15	24	*	*	100	242
2,000 or more	10	48	18	24	*	*	100	324
Single plant firm								
Under 50 employees	3	37	8	19	6	27	100	730
50–499	4	52	15	28	*	1	100	213
500 or more	6	54	12	26	*	2	100	68
Change in size of firm								
More employees	8	46	17	25	1	3	100	1,170
Same	4	46	7	17	3	23	100	774
Fewer	9	44	12	17	5	13	100	283
Don't know	3	34	21	42	*	*	100	352

*Less than 0.5 percent.

[a]Number of employees in the respondent's plant or office.

period of time by the growth of the firm's market. At least this was true during the particular 5-year period studied (on the average the machine change may have preceded the interview by about two and a half years).

This analysis of the kinds of workers who experienced changes in machine technology points to a number of characteristics which must cushion the impact of technological advance on the work force. Labor force members whose job was affected by equipment change are somewhat younger and better educated than the average, both factors which should aid them to make the needed adjustments. They also are more likely than others to work for large or expanding firms, a fact which should make it easier for their employers to retain them on the payroll and to find suitable work for them. Yet, it is also true that these relationships are not strong. The major conclusion to be drawn from the analysis just completed is that workers who experienced changes in machine technology have very diverse personal and employment characteristics. A significant minority does not have the good fortune, when technological change occurs, to be young, well educated, or to be working for a large or expanding firm.

Chapter 4

THE PROCESS OF CHANGE-OVER
TO THE NEW EQUIPMENT

When new equipment is installed in an office or plant, some workers may be laid off. Pay may be raised or lowered. Workers may or may not be transferred to another department. In addition, established work routines may be disrupted. Not only do workers have to learn how to operate the new equipment, often they are assigned new tasks. In this chapter we shall discuss these and other aspects of the change-over to the new equipment. We shall also be interested how (according to the workers' reports) the company handled the change-over. How much advance notice did the company give? Were the workers given a chance to participate in planning the change-over? What arrangements for training were made?

A distinction is made between the problems of the transition and longer-run consequences of the machine change for the worker. A man may experience unemployment and uncertainty about his future while the company installs new equipment. Yet after some months, when he has become skilled in his new job, he may advance more quickly than he could have with his old work. The opposite may also happen: A man may have no transitional problems, but in the longer run, he may not find his new job satisfactory. Therefore separate treatment of the two phases of adjustment is convenient. One limitation of our analysis should be recognized at the outset. Workers were not interviewed *during* the change-over but months, or even years, later. Therefore this chapter is confined largely to fairly objective matters which can be recalled reasonably well. No attempt is made to reconstruct the workers' frame of mind during the transition - his hopes, fears, or uncertainties.[1]

We shall be concerned in this chapter exclusively with the transitional phase. The longer run impact of changes in machine technology on the worker's economic welfare and job satisfaction will be analyzed in Chapters 5 to 7.

A. Advance Planning

It is generally stressed by industrial psychologists and management experts that communication and participation can help workers greatly in making an adjustment to new work situations.[2] This implies that workers should be given substantial advance notice when the company plans to install new equipment and should be acquainted with the company's plans for effecting the transition. Unions have emphasized this point.[3] Therefore all those who experienced machine changes which significantly altered their work were asked: "How much notice did your employer give you about the change he was making in machinery or equipment?" The answers show that workers for whom the machinery change did not involve a job change or transfer (column 1 of Table 4-1) often had very little advance warning. Over 35 percent of those who experienced significant changes in equipment heard about the prospective change less than 3 days before the new machinery was installed. One-half heard 2 weeks or less ahead of time. Only about 40 percent of workers were informed 3 months or more in advance. Nevertheless, when asked, "Do you feel that you had enough advance notice or was it insufficient?" over 85 percent of this group of workers said that they had been given enough advance notice. There seem to be several explanations for the infrequency of the answer "insufficient." For one thing, the group to which the data in column 1 refer could remain in the same department or section of their company. In retrospect (the interview often occurred 2 to 5 years after the equipment change) the problem of advance notice may seem rather inconsequential. More important, it is quite clear from the interviews that the new equipment often was so superior to the old from the workers' point of view

---

[1]See Charles R. Walker, *Toward the Automatic Factory: A Case Study of Men and Machines,* New Haven, Yale University Press, 1957, pp. 26-106, where a longitudinal study reveals the pattern of these transitional attitudes. Original fear and mistrust of automated machinery are gradually replaced by confidence in the equipment and the workers' ability to effectively run it.

[2]See for example Floyd C. Mann and L. Richard Hoffman, *Automation and the Worker,* A Study of Social Change in Power Plants, Henry Holt and Company, New York, 1960, pp. 191-202.

[3]See AFL-CIO, "Adjusting to Automation," Publication No.144, Washington, D. C., 1969, pp. 7-11.

TABLE 4-1

PLANNING THE CHANGE-OVER: ADVANCE NOTICE OF MACHINE CHANGE

	Machine change[a] and	
	Same job for past 5 years	Left old job because of machine change
Amount of notice		
2 days or less or none	37%	14%
3 days to 2 weeks	11	10
3 weeks to 2 months	11	14
3 months or more	41	62
Total	100%	100%
Adequacy of notice		
Enough	87%	82%
Pro-con	2	*
Not enough	11	18
Total	100%	100%
Number of cases	(170)	(32)

*Less than 0.5 percent.

[a]Excludes self-employed workers. Those for whom the information concerning notice was not ascertained are excluded from the percentages.

that they felt pleased to receive it. They knew that the new machines would make their work easier in some important respect. They spoke of new equipment the company "gave them" rather than new equipment they "had to use." Obviously if someone is about to do you a favor, he does not have to give advance notice.

Column 2 of Table 4-1 relates to the small group which had a more difficult adjustment to make—they left their employer because of the equipment change, either voluntarily or involuntarily. This group (column 2) did receive more advance notice. Yet even in these cases nearly a quarter reported 2 week's notice or less; but again only about one in six thought they had received inadequate advance warning.

Some of the diverse attitudes toward advance notice:

A 46-year old baker: Recently the large university for which he works introduced automatic ovens and large automatic mixers. The mixers greatly reduced the physical effort required for the job. Practically no notice of the change was given, although some men had to be shifted to other work. Yet the respondent did not complain. "It was planned by the university, but it was planned for our benefit too. It made our work easier."

A geologist in a government laboratory: Three months ago the lab received a computer that reads maps. This made his work easier, especially reduced the paper work. He had 2 weeks advance notice and feels that this was sufficient. "There had been a lot of begging for this computer for a long time."

A 26-year old former steel worker argued that "notice makes no difference. They [automated machines] still make an interesting job boring. Welding is interesting; it's boring to watch a machine weld." He has gone back to college after 1½ years with the steel company.

A 20-year old woman telephone operator, who went through the change-over to direct distance dialing a year ago complained about insufficient notice—one month, "I like to know more about what is going on and not be in the dark about it."

A 42-year old man who helps to schedule production in an automotive plant: Two years ago the reporting and scheduling operation was computerized. He complained that the company gave insufficient advance notice. "We just dove into it. They don't give you enough knowledge; and you have the feeling your job is going to be taken away."

A second feature of the change-over which is often stressed by industrial psychologists is cooperation between management and the workers in planning the change-over to new equipment. Accordingly, people who experienced significant changes in machinery were asked "Was the change-over planned by the company alone, or did employees participate in the planning, or did a union have some say also, or what?" The figures in the following table show that about two-thirds of those who experienced changes in machine technology reported that the company alone planned the change-over. Unions were rarely given any credit for participation in planning. If in fact they did in some instances play a significant role, the rank and file seems to have been unaware of their contribution. Those who did not see the company as doing all the planning most frequently said that the transition was planned jointly by the company and employees. In a few cases the respondents even asserted that the change-over was planned by the employees alone. Answers to that effect were occasionally given by people in supervisory positions who were referring to employees at a middle management level. It was also given, however, by employees who suggested to their boss that their work could be made much easier or be performed better if he acquired a new piece of equipment of which they were aware.

The following case illustrates this situation:

WHO PLANNED THE CHANGE TO NEW MACHINERY?

Company alone	66%
Employees alone	7
Other	4
Company and union	3
Company and employees	14
Company and other	3
Company, union and other	3
Total	100%
Number of cases[a]	(202)

[a]Includes those who have changed machinery but have not changed jobs and those who have changed jobs because of a machinery change. Those for whom the planning information was not ascertained are excluded from the percentages.

A 47-year old bookkeeper (income $5,000-6,000) for a real estate company, speaking of the acquisition of a bookkeeping machine: "I participated and planned it. I investigated it, picked it out. I got what I wanted. I set it up, and they went along with it. The first year was difficult in some respects to get it set up completely; but now my boss appreciates it as much as I do. I was already trained to operate it. I had used one in another office before I took this job." This machine replaced two other bookkeepers.

B. Training

Advance notice and planning, when new equipment is installed, may be important (among other reasons) to permit the worker to acquire needed training. People who experienced changes in machine technology which affected their work significantly were questioned about the training they received at the time of the change-over. Altogether, extensive retraining was infrequent.[4] Slightly over 40 percent of the group answered that no training was needed, either because the operation of the new equipment was very simple or because it was similar to the operation of previously used equipment (Table 4-2). Another 10 percent explained that they had to learn by themselves on the job how to work with the new equipment.

---

[4]This observation is supported by the investigations of James R. Bright, "Does Automation Raise Skill Requirements?" *Harvard Business Review,* July-August 1958, pp. 85-98. However, the 13 plants studied by Bright were automated in the strict sense of the term, while a much wider range of technological changes was experienced by the sample studied here.

TABLE 4-2

TRAINING AT THE TIME OF THE CHANGE-OVER TO NEW MACHINERY

		Machine change and		
			Left old job for other reasons	
Type of training	Same job for past 5 years	Left old job because of machine change	Transfer	Different employer
Formal course and on-the-job[a]	4%	18%	6%	4%
Formal course[a]	7	12	6	5
On-the-job	26	23	43	43
Self-training only	10	6	10	10
No need for training	53	41	35	38
Total	100%	100%	100%	100%
Number of cases	(170)	(32)	(49)	(282)

[a]Includes those who may have reported some self-training in addition to the training listed.

The questions asked were: "In order to work with the new equipment, did you have to learn anything new or did you acquire any new skills?" "How did you acquire the new skill or knowledge--did you learn it by yourself on the job? Did someone train you on the job? Or did you take a formal training program or course?"

On-the-job training was much more frequent than formal training programs. About 40 percent of those who experienced a significant machine change were given on-the-job training. On-the-job training programs varied greatly in length. The median length was 25 to 40 hours; but there were a good many reports of programs which lasted 8 hours or less as well as of others which lasted more than 160 hours.

Finally, somewhat over 10 percent of those who experienced machine changes took a formal training program, either by itself or in conjunction with on-the-job training. Most often the formal training program was given by the employer; but in a substantial proportion of cases it was given by a school or university. Occasionally it was given by the company which sold the equipment. More often than not, the training program was voluntary. In a small number of instances the worker or the government paid for it; but usually the cost was borne by the employer. In the majority of cases the formal training program was taken *after* the new equipment was installed, although occasionally the training program was part of the advance preparation. In about half the cases the formal training program extended over more than 160 hours; it seldom lasted less than 25 hours.

At the end of the sequence about training for the new equipment, people were asked "Altogether what helped you most in getting used to the change in your work? Was there anything which made it hard for you to get used to the change? What was it?" Placed as they were, these questions gave people an opportunity to refer to the usefulness of their training or to complain about its inadequacy. In all, many more workers could think of something which helped them to get used to the change in their work than could think of something which made it hard for them (Table 4-3). Only about 14 percent of the group with significant changes in machinery said that their training helped them; not even 2 percent referred to lack of training as the chief handicap in adapting to the new equipment. Many workers said that what helped them most were practice and time.[5] Other frequent references were to the helpful attitude of fellow workers and previous experience with similar equipment. Difficulties cited most often related to the nature and design of the new equipment. The supervisor was criticized with some frequency.

Education and training in relation to technological change will be explored further in Chapter 8.

C. Transitional Unemployment

Of particular concern is the impact of the new machines on employment and the incidence of unemployment during the change-over. In most of the discussion in this section we shall disregard the self-employed, since a man who owns his own business ordinarily is not subject to unemployment.[6] We shall also disregard people who changed jobs or were transferred, unless the job change or transfer occurred (according to the respondent) *because of* a change in machine technology. Our interest then centers on the unemployment experience during the change-over of workers who experienced a significant change in machine technology during the past 5 years *and* (a) worked for the same employer during the entire period or (b) were transferred or changed jobs *because of* a change in machine technology. This group represents a small part of the sample (7.6 percent of the labor force or 202 cases in our sample). The data should therefore not be read too closely; they are indicative of orders of magnitude. Nevertheless they seem to tell an unambiguous story.

---

[5]Walker also found that getting used to the new equipment contributed most to changing negative attitudes to positive ones, *op. cit.,* p. 141.

[6]People who were self-employed were asked whether they had worked for someone else in the past 5 years, and if so, whether they left because he changed to new machinery or because of automation. It happened that there was no such case in the sample.

TABLE 4-3

FACTORS WHICH WERE OF THE MOST HELP OR CAUSED THE MOST DIFFICULTY
IN ADAPTING TO NEW MACHINERY

| | | Machine change and | | |
| | | | Left old job for other reasons | |
Mentioned as helpful	Same job for past 5 years	Left old job because of machine change	Transfer	Different employer
Training and education	12%	13%	16%	16%
Previous experience with similar equipment	8	6	13	14
Practice, time	23	6	13	20
Help from supervisor	4	3	7	8
Help from other workers	*	10	18	16
Design of equipment; other	6	13	13	13
Nothing helpful mentioned	47	49	20	13
Total	100%	100%	100%	100%
Mentioned as difficulties				
Insufficient training and education	1%	*	4%	1%
Insufficient practice, time, or experience	2	3%	4	2
Lack of help from supervisor	3	*	4	4
Design of equipment; other	7	26	25	19
No difficulties mentioned	87	71	63	74
Total	100%	100%	100%	100%
Number of cases	(170)	(32)	(49)	(282)

*Less than 0.5 percent.

In Chapter 3 (Table 3-3) it was found that in somewhat less than half the cases where respondents reported significant changes in machine technology, they estimated that after the machine change fewer workers were needed to do the *same amount* of work per day. However, companies often order new and more efficient machines when the market for their product is expanding. With the volume of work increasing, the installation of a labor-saving machine may not require layoffs or transfers. Frequently also, companies introduce labor-saving machinery when they have been facing a labor shortage and have been paying for overtime work. With these possibilities in mind, a further question was asked referring to events after the change-over:

"Did the change-over to the new equipment affect the number of people needed to do the work *in your section*? Were more or fewer people needed after the change-over?"

By far the most frequent answer, given by over 70 percent of the group under study here, was that there was no effect on employment in their section (Table 4-4). Another 21 percent reported that employment was reduced, and less than 10 percent that employment increased. It would appear then that in at least half of the cases where labor-saving machinery was introduced, employment in the respondent's own section was not reduced.

We know that a given reduction in the work force does not mean an equivalent number of layoffs: some workers may be transferred; some retirements or resignations are bound to occur for reasons unrelated to technological change. Thus, although 21 percent of the group reported a reduction in employment in their section, only 18 percent were aware of any layoffs or shorter hours in their section during the transition. Shorter hours were less frequent than layoffs. When there were shorter hours or layoffs, usually "quite a few" people were reported to be involved. A third of the subgroup who were transferred or changed jobs *because of* a change in machine technology said there had been unemployment in their section.

Next, people were asked "Was your own job abolished by the new equipment?" This question appears to have had an ambiguous meaning to respondents and may therefore have produced some understatement. Job content is often altered when different equipment is introduced. It then becomes difficult to draw the boundary line between a situation where the old job was abolished and the worker given a different job in the same section and a situation where the job continued, although involving rather different duties. Those who remained in the same section said in a few instances that their former job was *partly* abolished; but overwhelmingly their continuance in the same section led them to feel that their job was *not* abolished. In all, 10 percent of the group under study reported that their job was abolished as a result of a change in machine technology. Practically all these people belonged to the subgroup who were transferred or changed jobs *because of* a change in machine technology. Half of this same subgroup reported however that their job was not abolished. In some cases the new work did not suit them, the hours were no longer satisfactory, the employer needed them in another section where new equipment was being introduced, and the like.

The report "My job was abolished" is different from the report "I myself was laid off." A man may be laid off temporarily during the change-over, even though his job will later continue. On the other hand, as we saw, some

TABLE 4-4

EMPLOYMENT EFFECTS OF MACHINE CHANGE

Change in the Number of People Needed in Section

Fewer	21%
Same, not affected	72
More	7
Total	100%

Whether Anyone in Section was Unemployed or Working Shorter Hours

Yes, unemployed	13%
Yes, shorter hours	4
Yes, both	1
No	82
Total	100%

Whether Worker's Own Job was Abolished

Yes	10%
Part of job	3
No	87
Total	100%

Whether Worker was Himself Unemployed,
Working Shorter Hours, or Stopped Working

Yes, unemployed		4%
Yes, shorter hours		2
Stopped working		3
To look for a better or more secure job	1	
Personal reasons, not ascertained why stopped working	2	
Longer hours		3
No, nothing like that happened		88
Total		100%
Number of cases		(202)

of the people whose job was abolished were transferred or obtained a new job without intervening unemployment. Among the group under study (i.e. those who experienced technological change but no job change and those who had a job change or transfer *because of* a change in machine technology) roughly 4 percent reported that they themselves were unemployed during the transition and another 2 percent that they were working shorter hours; a few others quit their job. Three percent were working longer hours during the change-over.

There is little doubt about the conclusions to be drawn from these data, however rough they may be. Although we started out with an estimate that during the past 5 years nearly half of the reported changes in machine technology which had a significant impact on people's work may have been labor-saving, it turned out that only 4 percent of the workers reporting changes became unemployed and another 2 percent had shorter hours. Six percent of the group under study here constitute barely one-half percent of the labor force. This figure may be somewhat of an underestimate in that it disregards unemployment of job changers who said that their employment shift had nothing to do with machine change. Technological change may have been involved in a few of these cases in ways which were not evident to the respondent.[7] In any case, the order of magnitude of the unemployment estimate would remain similar. Thus, unemployment which stems *directly* from changes in machine technology seems to have been very infrequent in the period studied. The low incidence of *direct* unemployment is all the more striking since 21 percent of the entire sample reported some unemployment during the same 5-year period.[8]

The reason for the large gap between the frequency with which labor-saving machinery was reported and the frequency with which unemployment for the respondent resulted directly from such machine change is evident from our analysis. First, in an expanding market, labor-saving machinery need not imply a reduction in the work force. This is especially true if, as we have seen, growing firms are more likely than others to introduce new equipment. Secondly, a reduction in the work force may be achieved by normal resignations and retirements, without any need for layoffs or shorter hours. Thirdly, even when layoffs become necessary, only a fraction of the people working with the labor-saving equipment are affected. And finally, in a growing economy, job changes and transfers absorb some of the workers who might potentially be laid off, minimizing transitional unemployment. The residual fraction of workers whose unemployment can be traced *directly* to labor-saving technological change was very small indeed during the period under

---

[7]See the discussion of this issue in Chapter 3, Section B.
[8]For a definition of the concept of "direct unemployment" as used here see Chapter I, Section D.

study. This analysis does *not* contradict the proposition that labor-replacing technological change may reduce the potential volume of employment. It permits us to infer that whatever unemployment results from technological change must be very largely *indirect:* Someone who might have been hired is not hired. Perhaps a new entrant into the labor market must wait longer to find work; perhaps someone who quits his job or is laid off for reasons unrelated to technological change must search longer until he finds a new one.

If this conclusion is correct, it follows that technological change seldom leads to unemployment for an experienced skilled worker with a satisfactory work record, even though a machine may take over his former duties. Rather, technological change has its major impact indirectly on the most marginal groups in the labor market: those at the end of the employment queue. These may be the very young and the oldest age groups, the least educated and skilled, Negroes, and those who may be handicapped for other reasons in finding work. Technological unemployment seems to trickle down to these groups to a large extent.[9] To say that technological unemployment among skilled workers in a strong labor market position is rare in periods such as 1962-67 is not to deny that it should be a concern of public policy. Even if only one-half to one percent of the work force were *directly* affected by automation over a period of 5 years, these unimpressive percentages would represent 350,000 to 700,000 human beings. Moreover, some skilled workers may avoid unemployment by accepting work which offers them less satisfaction than their former job. Chapters 6 and 7 will throw more light on this possibility.

In this section we were concerned with unemployment at the time of change-over to the new machinery. In the next chapter we shall look at the longer-run employment record of those who experienced changes in machine technology. Until then the conclusions advanced here remain tentative.

D. Pay Changes

One indication of the extent to which a person's job was upgraded or downgraded at the time new machinery was introduced is change in pay, seniority rights, and fringe benefits at the time of the change-over. Respondents were asked: "Was your pay raised by the introduction of the new equip-

---

[9]It has frequently been discussed whether income growth "trickles down" sufficiently to be counted upon to reduce poverty. See W. H. Locke Anderson, "Trickling Down: The Relationship Between Economic Growth and the Extent of Proverty Among American Families," *Quarterly Journal of Economics,* November 1964, pp. 511-524. The hypothesis advanced here—that technological unemployment *does* trickle down to the weakest group in the labor force would, if correct, reinforce the view that poverty is not readily abolished by more vigorous economic growth, to the extent that such growth resulted from accelerated technological change.

TABLE 4-5

CHANGES IN INCOME, SENIORITY RIGHTS, AND FRINGE BENEFITS
AT THE TIME OF MACHINE CHANGE

| | | Machine change and | | | No machine change and | |
| | Same job for past 5 years | Left old job because of machine change | Left old job for other reasons | | | |
			Transfer	Different employer	Transfer	Different employer
Income change						
Increase	16%	39%	53%	55%	54%	56%
No change	82	26	35	19	40	25
Decrease	2	35	12	26	6	19
Total	100%	100%	100%	100%	100%	100%
Change in seniority rights or fringe benefits						
Improved	1%	10%	15%	23%	15%	16%
No changes	98	66	76	51	81	59
Reduced	1	24	9	26	4	25
Total	100%	100%	100%	100%	100%	100%
Number of cases	(170)	(32)	(49)	(282)	(160)	(466)

ment (when you changed jobs), or lowered, or was it unaffected?" By this criterion, upgrading accompanied technological change much more frequently than did downgrading, although there were differences among groups, and many people reported no change. Among wage and salary earners for whom the changes in machine technology implied no job change or transfer, pay rates or earnings usually were not affected by the introduction of new equipment. Only 2 percent of the group reported that their pay was lowered, while about 16 percent said that their pay was raised at the time of the change-over. The remaining 82 percent experienced no change in pay at that time (Table 4-5, column 1). The "no change" group includes a number of workers whose hours were reduced or overtime eliminated by the greater efficiency of the new machines, but whose pay rates were raised to compensate them for the loss. Fringe benefits and seniority rights were affected very rarely in these cases.

Many more income increases as well as decreases in pay occurred among those who were transferred and changed jobs. In general, transfers often imply a promotion and are associated with raises in pay. Similarly, to obtain a higher wage or salary is a major motivation for changing jobs. Thus over one–half of the job changers and transfers reported that their pay was increased when they made the shift. The one exception is the group who changed jobs or were transferred *because of technological change* (column 2). Some of these people were transferred or moved to a job which was newly created by technological advance and for which their skills or experience particularly qualified them. Such a change usually meant an increase in pay. However, many others in this category were displaced by automation of their former job and therefore tended to be in a weak bargaining position. In consequence a smaller proportion of this group than of other job changers and transfers received higher pay when they changed jobs. Decreases in pay at the time of the change-over are more frequent among job changers than among transfers. They are more frequent among workers who had to use different equipment on the new job than among those who did not have to make such a change. And finally, decreases in pay are most frequent among those who changed jobs or were transferred *because of* technological change.

Fringe benefits and seniority rights seldom were reduced by transfers; but about one fourth of workers changing employers reported reductions. Many other job changers were young and had no seniority rights to speak of.

E. Conclusion

The statistical data presented in this chapter show quite clearly that only a small minority of people suffered economic setbacks during the change-over to new equipment. In the prosperous years 1962-67 few people needed to be dismissed because of automation; and those who did lose their job often found another without any intervening period of unemployment. Downgrading of the job, and hence of pay rates, took place in about a third of cases where workers sought a transfer or new employment *because of* technological change. However, technological change infrequently precipitated such employment shifts.

A reading of the interviews with people who experienced significant changes in machine technology confirms the conclusions conveyed by the statistical data. It is almost impossible to find any cases which suffered *serious* hardship during the change-over in the period 1962-67. In order to convey a qualitative picture of the kinds of problems which some people did encounter, a few cases are summarized below. It should be emphasized that these cases are *not* representative of all workers who underwent a change in machine technology. They are typical of those few people with unfavorable experiences.

*   *   *

A 51-year old mechanic working for a copper company. His job is to repair heavy equipment trucks (income $7,500-9,999): Three years ago he was a traffic control man on the company's railroad haulage system. Then the company shifted some of its haulage from railroad to truck. At the same time it changed from manual to electric switches. A hundred or more jobs were abolished, including his own. He knew in advance and applied for every opening in the company that came along. These were filled on the basis of seniority. He was transferred without intervening unemployment. However, his pay and seniority rights were lowered. "I started over again at the bottom in the mechanical department." He likes the new job better than the old. "There is a greater variety of things to do and to learn and there is a challenge. I enjoy learning."

A 36-year old steel worker (income $6,000-7,499): "I slit the steel coils when they come out. I slit them into different sizes. I push a button, and it cuts the steel coil at a certain length." Three years ago his company changed from a 10-inch mill to a 22-inch mill. This meant that quite a few people were laid off. He did not become unemployed, but is afraid that with further automation his job will be abolished. "I got bumped down in pay scale. I mean my top was lowered."

A 59-year old married woman, clerk-stenographer with a state Bureau of Parole: She had been on her previous clerical job for 9 years. They installed an automatic machine to put invoices on tape. "When the new machine came in, it was nothing but total confusion. More people were needed because there was so much more paper work - so many copies to be filed. I stopped working. There was so much additional work, I got a nervous breakdown." She was out of work for 1½ years and then accepted the job with the Parole Bureau at a lower salary.

A 38-year old female worker in a cigarette factory, Negro: Three years ago more automatic equipment was introduced and the work force was reduced. In addition, those who remained were put on shorter hours. She received a raise but reports that her earnings are lower than 5 years ago, since she now works 7½ hours instead of the former 9 hours a day.

A former elevator operator in a government building, age 52, Negro; 9 years of schooling: He lost his job 1½ years ago when the government installed automatic elevators. "Those of us who didn't have any military years behind us were laid off." He was out of work for one month and then got a job as janitor in a department store at a lower salary. He now makes $3,000-3,999, but says he likes his new job better than the former one.

A 41-year old tool machinist in a tractor factory, born in Poland, 6 years of schooling plus 3 years of vocational training: Two and a half years ago he was a machine repair man and also made replacement parts. Then the company introduced new machinery for which spare parts could be bought from the manufacturer. As a result, the repair crew was reduced in size. He lost his job. The company offered him a job at a lower pay rate, but he refused it. After 3 months he was recalled and transferred to his present job with no loss in pay. During the intervening 3 months he worked for another company. He finds his present job more routine than the old one, since he now makes machine parts according to standard specifications. His income is $7,500-9,999.

A 32-year old secretary with a large chemical company, married: She used to be a bookkeeper and accountant for a retail chain. Six months ago the retail chain computerized its accounting system and dismissed a number of people in that department; they also transferred her boss. She disliked the new machines and the new boss. Her own job was not abolished but "over-loaded." "The way I saw it, we needed more people until we could get set up better. I could not agree with the new boss, so was fired.   They asked me to resign, but I wouldn't."   There was no un-employment. "I went right to work where I am now." The new job pays less than the old one.

Chapter 5

ECONOMIC CONSEQUENCES OF CHANGE IN MACHINE TECHNOLOGY

The purpose of the chapter is to examine the effects of machine use and machine change on the economic position of the worker. Chapter 4 concerned itself with the immediate, and sometimes temporary, effects on the worker of a change to new equipment. This chapter is concerned with the longer-run economic implications of equipment change.

The chapter is focused therefore on the 5 years immediately previous to the survey. The dependent variables studied— income change, promotions, and unemployment—refer in general to the same 5-year period. However, the machine change may have occurred at any time within the 5-year period; there may have been more than one change or even occasionally a continuous process of change. It was not practicable to have workers date these changes. This means that recent changes in machinery will have had less time to affect economic variables than changes occurring 4 or 5 years ago. We might then assume that the effects of machine change would be understated (or overstated) to the extent that longer-run changes in economic variables are not fully worked out. Moreover, some income changes or unemployment may have occurred before, not after, the machine change and thus may lack any relation to the change in equipment. It is unlikely that these problems will distort the comparisons seriously. Still, in order to circumvent this difficulty, additional questions were asked about expected future changes in income and other economic variables. The expectational measures may reflect the perceived *consequences* of recent machine change more fully than do questions referring to past economic changes.

This chapter is largely the work of Charles P. Staelin.

Section A deals with the relation between the *level* of machine technology and the economic position and progress of the worker. It examines the association of equipment use and automation level of equipment important to workers' jobs, with income level, past and expected income change, and unemployment. Section B is concerned with the relation between *changes* in machines and machine technology, and these same economic variables. Both Sections A and B suggest that the direct economic impact of machine technology is largely beneficial. Section C presents a multivariate analysis of the relations examined in the first two sections. This analysis demonstrates that even after socio-economic variables have been accounted for, no major adverse economic effects of technological change on those workers directly involved in the change can be detected. The chapter is concluded with Section D, a brief discussion of the effects of machine use and machine change on the job mobility and geographic mobility of workers.

A. Machine Use and Economic Situation of the Worker

Although the main subject of this study is the relation between machine *change* and the economic progress of the labor force, a brief look at the *level* of machine technology is warranted here as a starting point for an analysis of machine change. This section will explore the associations between equipment use and the automation level of workers' equipment on the one hand and their economic situation in the past 5 years on the other hand—promotions, past and expected income change, unemployment experience, and income level.

It should be emphasized that we can examine only the *direct* effects of machine use and the level of machine technology on a worker's economic position. Yet the use of sophisticated production equipment may not only raise the income of its operator, but it may raise the firm's profits and allow it to pay all other workers higher wages. Our analysis identifies the contribution of the machine to the economic situation of only those workers who reported that it was important to their job. In a perfectly competitive world there should be no relation between the automation level of a worker's equipment and the level of his income, other things being the same.[1] The contribution to output of workers with the same job qualifications should be identical; otherwise the firm would hire more of those workers with higher contributions to output and hire fewer of those with lower contributions to output.

---

[1]This idea is discussed in more detail by W. Allen Wallis, "Some Economic Considerations," in The American Assembly, *Automation and Technological Change*, John T. Dunlop, ed., Prentice-Hall, 1962, pp. 108-110.

However, this type of long-run textbook equilibrium is seldom achieved. People doing the same kind of work are not all equal in their talents and skills. Wide diversity occurs in education, among other things. Also, the labor market is far from being perfectly competitive. The labor requirements associated with a given machine set-up are fairly rigid. Finally, the long period of time needed for adjustments toward the theoretical equilibrium seldom materializes. Changes in the economy come so quickly upon one another that the chances for a complete adjustment to any one change are negligible. The differences in incomes between workers are therefore in part reflections of shorter-run advantages or disadvantages resulting from the equipment they work with. Table 2-4 did show some earning differences between the various machine-use groups, but these earning differences were also associated with occupational and educational differences which could readily account for them.

Table 5-1 uses the Automation Scale introduced in Chapter 2; added are those workers who do not operate machinery at all. The table shows the relation of automation level to workers' incomes from their main jobs for the year 1966. This is before-tax income and does not include any income which the worker may have received from secondary jobs, from other family members, or from other sources, such as financial assets. The median income falls from $8,200 for those in the highest automation group to $5,400 for those operating the least mechanized equipment. Although the differences in income among the lower three automation levels are neither large nor consistent, it is clear that income does tend to be higher for those operating logically controlled equipment. The characteristics of workers using various types of equipment have been discussed in Chapter 2. It was noted there that those members of the labor force who use more mechanized or automated equipment are appreciably better educated and tend toward the more professional and highly skilled occupations. It is not surprising therefore to find that the relation between the automation level of the equipment a worker operates and the income level of the worker is positive. Chapter 2 further showed that those workers with no equipment tended to be of two types, one poorly educated and concentrated in the unskilled and service occupations, and one fairly highly educated and concentrated in the upper occupational groups. The income figures reflect this. Those with no equipment have, with above average frequency, incomes below $2,000 and above $10,000, while they are less often in the middle income brackets. The dual nature of the group without equipment is reflected in all the following tables. It does not, however, obscure the increase in economic status associated with the use of very sophisticated equipment.

The lower part of Table 5-1 indicates that people whose work is affected indirectly by equipment have higher incomes than those who reported no indirect contact with machines. And again, those who have indirect contact

TABLE 5-1

WORKERS' INCOME AND THE AUTOMATION LEVEL OF EQUIPMENT OPERATED DIRECTLY AND USED INDIRECTLY

1966 Income from worker's job	Automation level of equipment operated directly				
	Numerical, tape, computer or other logical control	Fixed mechanical control	Powered multi-system, manual main control	Manual control, operator powered or powered single-system	No equipment
Less than $2,000	3%	11%	10%	14%	12%
$2,000-2,999	3	9	9	8	8
$3,000-3,999	5	14	11	11	9
$4,000-4,999	5	12	10	11	7
$5,000-5,999	13	12	13	13	9
$6,000-7,499	16	14	18	14	13
$7,500-9,999	22	15	13	14	15
$10,000-14,999	22	8	10	11	15
$15,000-24,999	8	2	3	1	7
$25,000 or more	3	1	2	*	4
Not ascertained	*	2	1	3	1
Total	100%	100%	100%	100%	100%
Median income	$8,200	$5,300	$5,800	$5,400	$6,600
Number of cases	106	1,038	409	348	743
	Automation level of equipment used indirectly				
Median income	$8,900	$6,800	$6,800	$7,700	$5,200
Number of cases	173	405	157	42	1,857

*Less than 0.5 percent.

with computers and other logically controlled equipment reported the highest earnings on the average.

Another important aspect of economic welfare is change in income. The time path of a worker's income is often more significant than its present level; in the next chapter income change will be shown to have a considerable effect upon the worker's attitude toward automation. Three variables have been chosen here to measure change in income. They are the direction of the movement in a worker's income from his main job over the past 5 years,[2] the occurrence of one or more promotions during the past 5 years, and the expected direction and magnitude of future income changes. The wording of the questions appears at the bottom of Table 5-2.

Income change seems to have little, if any, relation to machine use. In the labor force as a whole 73 percent of people reported that they had increases in the income obtained from their job during the past 5 years. Workers whose job has no significant connection with machines at all had the lowest incidence of income increases (69 percent) and the highest frequency of no change (15 percent). For people who operate machines but have no other contact with equipment, the percentage was 71. The group which benefitted most frequently from income increases was that whose work is affected by equipment they do not operate (80 percent) and those who operate *and* have indirect contact with machines (78 percent). Reported promotions during the past 5 years, though much less frequent than income advances, follow the same pattern. Their frequency rises from only 27 percent among machine operators, to about 40 percent among those who have indirect contact with machines (regardless of whether they do or do not operate machines in addition). These differences are not impressive, and they may be partly due to other factors. We saw earlier that those who work indirectly with machines are also the best educated members of the labor force.

Table 5-2 reveals that income increases, both past and expected, as well as the frequency of promotions do rise with the technical level of the worker's equipment. The occurrence of promotions over the past 5 years rises most markedly from 23 percent at the lowest automation level to 62 percent at the top.[3] Those with no equipment have a slightly higher incidence of promotions than those who operate low automation level equipment, due possibly to presence of a highly educated segment in the former group. However those with highly automated equipment do far better than those with

---

[2]The reader should note that the income change variable analyzed in Chapter 4 measured the incidence of raises or pay cuts *at the time of the change-over* to the new equipment and/or job. The income change analyzed here relates to the past 5 years.

[3]Part of this disparity is due to the fact that promotions are a phenomenon occurring primarily among white collar workers. These workers are more concentrated in the higher automation level groups.

TABLE 5-2

PROMOTIONS, AND PAST AND EXPECTED INCOME CHANGE
AND THE AUTOMATION LEVEL OF EQUIPMENT OPERATED DIRECTLY

	Automation level				
	Numerical, tape, computer or other logical control	Fixed mechanical control	Powered multi-system, manual main control	Manual control, operator powered or powered single-system	Operate no equipment
A. Promoted - past 5 years					
Yes	62%	32%	26%	23%	31%
No	37	65	73	76	66
Not ascertained	1	3	1	1	3
Total	100%	100%	100%	100%	100%
B. Income change - past 5 years					
Increase	93%	81%	73%	73%	76%
No change	4	12	16	18	14
Decrease	2	5	10	7	7
Not ascertained	1	2	1	2	3
Total	100%	100%	100%	100%	100%
C. Expected income change					
Increase	92%	81%	71%	74%	64%
A lot	31	18	15	19	16
A little	61	63	56	55	48
No change	1	10	16	14	19
Decrease	1	12	3	3	7
Don't know; not ascertained	6	7	10	9	10
Total	100%	100%	100%	100%	100%

*Less than 0.5 percent.

The questions asked were: "Were you promoted at anytime during the past 5 years? Looking back 5 years, has your income from your job increased, stayed about the same, or has it gone down? And how about the next few years - do you expect that your income from your job will rise a good deal, rise slightly, remain the same, or go down?"

none. It is clear that those who operate more sophisticated equipment not only have higher incomes but also have a greater tendency toward rising incomes. The workers do not see this as a temporary phenomenon, for the expectations of those in the various automation groups follow the same pattern as past income change (Table 5-2).

Another question which is much discussed is the relationship between machinery use and unemployment. Respondents were asked whether they were ever unemployed for a week or more. Those who answered in the affirmative were then asked: "When was the last time you were unemployed for a week or more?" Table 5-3 shows that in the cross-section of labor force participants, 21 percent reported some unemployment in the past 5 years and 16 percent during the past 2 years. Irrespective of the time span examined, there is *no* evidence that the two categories of machine operators had more unemployment than those who do not work with machines at all. The unemployment percentages of these two groups are nearly the same. The group with the lowest unemployment is that which has only indirect contact with machines. Only about 6 percent of the labor force had more than two spells of unemployment in the 5 years under study. This proportion was highest among people who do not work with machines at all. It was marginally lower among all three kinds of machine users.

Table 5-4 indicates that highly automated equipment is associated not only with above-average advances in income but also with relatively low unemployment. The incidence of unemployment in the last 5 years for those who operate equipment falls from 25 percent to 13 percent as we move up the Automation Scale. Repetitive unemployment is also less frequent for those high on the Automation Scale.

The automation level of equipment not operated by the worker, but otherwise important to his job was generally found to be above the automation level of the equipment the worker himself operated. This is due at least in part to the fact that more automated equipment tends to have more linkages to other workers and departments than does less automated equipment. Also workers may have been particularly eager to tell the interviewer about equipment related to their jobs which is more sophisticated than their own. Given this relation, it is not surprising that we find the pattern of Tables 5-2 and 5-4 repeated, when workers are classified by automation level of equipment which affects their work indirectly. It was also found that the incidence of promotions and income increases rises with automation level of equipment with which the worker has indirect contact, while unemployment declines.

The above observations are largely descriptive. For although the correlations between automation level and various measures of economic welfare are pronounced, the technical level of a worker's equipment is also related to education and, to a lesser extent, to other personal characteristics. A mul-

TABLE 5-3

UNEMPLOYMENT AND WORKERS' RELATION TO EQUIPMENT

		Workers' relation to equipment			
Last period of unemployment[a]	All	Both operate and have indirect use	Operate only	Indirect use only	None
Within the last 5 years	21%	20%	23%	14%	20%
6 months ago or less; presently unemployed	7	6	8	3	8
7-18 months ago	6	5	7	4	5
2 years ago	3	3	3	3	2
3 years ago	2	3	2	2	2
4-5 years ago	3	3	3	2	3
Not in the last 5 years	78	79	76	86	79
Not ascertained	1	1	1	*	1
Total	100%	100%	100%	100%	100%

[a]A period of one week or more.

TABLE 5-4

UNEMPLOYMENT AND THE AUTOMATION LEVEL OF EQUIPMENT OPERATED DIRECTLY

	Automation level				
Number of periods of unemployment-past 5 years	Numerical, tape, computer or other logical control	Fixed mechanical control	Powered multi-system, manual main control	Manual control, operator powered or powered single-system	Operate no equipment
None in past 5 years	86%	79%	76%	72%	80%
One	8	11	11	10	9
Two	3	3	4	7	4
Three or more	2	5	7	8	5
Not ascertained	1	2	2	3	2
Total	100%	100%	100%	100%	100%

tivariate analysis which will be presented later in this chapter is designed to separate the influence of these factors. Yet it is difficult to interpret this statistical separation of influences. It seems to be inherent in more automated equipment that it requires workers with certain aptitudes and educational qualifications which would help a person to do well economically in any case.[4]

B. Machine Change and Change in Workers' Economic Situation

If the relation of machine level and use to economic status is somewhat difficult to interpret, the relation of machine *change* to the same variables is far less ambiguous. As noted in Chapter 3, machine change pervades all levels of the economy with no very strong association to occupation, education, age, or size of employer. It will appear throughout the following pages that equipment change does have some direct economic impact on the work force; the results are on balance favorable, although negative consequences are not wholly absent.

We are concerned here solely with machine changes which respondents considered as being of some importance for their work. Again, only the *direct* effects of equipment change on the incomes of those working with the changed equipment are studied. This section begins with a look at the relation between workers' 1966 gross income from their main job and the machine change which they have experienced during the past 5 years. Table 5-5 shows some significant and interesting income differences between the various machine change groups. Those workers experiencing machine change between mid-1962 and mid-1967 tended to have higher incomes in 1966 than those who did not, except when the machine change was the result of a job change.

For the work force as a whole the 1966 median income from the main job was $5,900. The group which had a change in machinery without a job change was concentrated in the middle and upper income brackets with a median income of $7,000 per year. By contrast, for the large group without job change *and* without machine change, the median income in 1966 was $5,600. This group includes those with no equipment use at all. Within the latter group the top and the bottom rungs of the economic ladder are heavily represented, as noted in the previous section. In looking at the self-employed, we again find that those with a machine change had generally higher incomes than those without a change.

---

[4]This point will be dealt with at greater length in Chapters 6 and 8.

TABLE 5-5

MACHINE CHANGE AND INCOME

1966 income from worker's job	Same job for past 5 years		Different job and[a]			Self-employed and	
	Machine change	No machine change	Machine change		No machine change	Machine change	No machine change
			Left old job because of machine change	Left old job for other reasons			
Less than $2,000	4%	13%	*	15%	10%	4%	8%
$2,000-2,999	3	8	6%	11	9	6	7
$3,000-3,999	6	11	19	14	12	9	9
$4,000-4,999	8	9	13	13	10	7	9
$5,000-5,999	14	11	3	11	13	11	8
$6,000-7,499	20	14	22	16	14	11	9
$7,500-9,999	21	15	28	12	16	13	11
$10,000-14,999	17	11	9	6	10	21	16
$15,000-24,999	5	3	*	1	3	11	10
$25,000 or more	1	2	*	1	1	7	9
Not ascertained	1	3	*	*	2	*	4
Total	100%	100%	100%	100%	100%	100%	100%
Median income	$7,000	$5,600	$6,600	$4,700	$5,700	$7,700	$7,200
Number of cases	(170)	(1,165)	(32)	(331)	(642)	(54)	(268)

* Less than 0.5 percent.
[a] Includes those who were transferred within the same company.

Job changers had lower than average incomes; they are also younger than average. Those who change jobs *and* equipment are particularly young (Table 3-6). Their median income was $4,700; the median income of job changers who used the same equipment before and after the job change was $5,700. It is interesting to note however, that those who had changed jobs or transferred *because* of a change in equipment had higher incomes in 1966 than other job changers. One reason may be that this group contained a higher proportion of transfers than the remaining group of job changers.

One must be cautious in attributing any causal relationship to the association of machine change with higher income level. Indeed any causation which might be present could run both ways. High labor costs might lead employers to search for labor-saving machinery; conversely, labor-saving machinery might tend to raise labor productivity and hence workers' incomes. In any case the positive relation between income and machine change indicates that, even if machine change dislocates some workers, machine change on the whole does not put the group which it affects into a disadvantageous economic position, at least in the context of a vigorously expanding economy.

More important than income level in relation to machine change is income change. For here we can postulate and test hypotheses more clearly than we could in considering income level. For those staying on the same job we might expect a tendency for incomes to rise with the introduction of new machinery, as the workers' productivity is likely to increase under such circumstances. Yet, in some cases workers might be displaced by machines which made their experiences and skills obsolete, and hence they might suffer economic reverses. We are interested in the relative frequency of such consequences. The incomes of the self-employed should almost certainly rise with the installation of new equipment, as entrepreneurs would hesitate to introduce different equipment unless they felt that it would benefit them financially. (Of course not all machine change undertaken by the self–employed is entirely discretionary; some is defensive or retaliatory.) The effect of new equipment on those workers who are changing jobs is less clear a priori. Most of those workers with a voluntary job change should have some increase in income as this is a prime motive for changing jobs. On the other hand some workers are forced to change jobs and may have to settle for lower wages in order to obtain a new job.[5] We should then expect that those workers changing jobs should show a high frequency of income increases *and* decreases and that an additional machine change might increase the frequency of both.

---

[5]The line between voluntary and involuntary job change is not a clear one. Some workers "voluntarily" change jobs when they experience unsteady work or short hours. A BLS study of job mobility found that in 1961 one-third of all job shifts were made to improve economic status. The study also noted that in a more prosperous year (such as 1955) this proportion is higher. See Gertrude Bancroft and Stuart Garfinkel, "Job Mobility in 1961," *Monthly Labor Review,* August 1963, pp. 897-906.

Many of these suppositions are borne out by the figures in Table 5-6. Workers experiencing a machine change within the past 5 years have in general had more promotions and more income increases than workers without a machine change. Looking first at workers who have had the same job for the past 5 years, the incidence of promotions is 35 percent and the incidence of income increases is 91 percent for those with a change in equipment; the corresponding figures are 23 percent and 83 percent for those without any change in equipment. Income decreases were infrequent among workers with equipment change, as they were among others.

For the self-employed machine change is similarly advantageous. As a group, the self-employed seem to have done more poorly with regard to income increases during the past 5 years than have wage and salary workers. It is conceivable that this group is simply more reluctant than others to report rising incomes.

Among those changing jobs, promotions and income increases occurred with about the same frequency whether or not there was a concurrent change in machinery. Contrary to expectations, job changers as a whole have had somewhat fewer income increases over a 5-year period than those without a job change. Apparently there were enough instances where the change in jobs was involuntary, or was preceded by an unfavorable income trend with the old employer, to bring about this result. Also, some of the unemployment associated with job change must have reduced 1966 income.

As an indication of the impact of a forced job change we can note the experience of those workers who left employers who were cutting back in production. They were consistently less well off than other job changers. Only 68 percent of those leaving an employer who was cutting back production had income advances; 28 percent had promotions. Of other workers who changed jobs 76 percent had income increases and 44 percent were promoted. As a further explanation of the comparatively low frequency of income increases among job changers, we should note that, even when job change is voluntary, there must be cases where income increases are temporarily sacrificed in return for greater chances of advancement or more job security. The latter explanation becomes plausible when we observe that workers who have changed jobs report more promotions than those who have not changed jobs. To be sure, the category "job change" in Table 5-6 includes transfers within the same company, which may often be promotions. Yet even disregarding workers with transfers, those with a job change reported more promotions than those without a job change.

The figures below suggest that the frequency of promotions and income increases is higher among those who have been transferred than it is among those who have changed employers.

TABLE 5-6

PROMOTIONS, PAST AND EXPECTED INCOME CHANGE AND MACHINE CHANGE

	Same job for past 5 years		Different job and[a]			Self-employed and	
	Machine change	No machine change	Machine change (Left old job because of machine change)	Left old job for other reasons	No machine change	Machine change	No machine change
Promoted - past 5 years							
Yes	35%	23%	56%	40%	40%	b	b
No	65	74	44	58	57	b	b
Not ascertained	*	3	*	2	3	b	b
Total	100%	100%	100%	100%	100%		
Income change - past 5 years							
Increase	91%	83%	73%	79%	76%	68%	51%
No change	7	12	14	10	13	28	31
Decrease	2	2	10	11	9	4	16
Not ascertained	*	3	3	*	2	*	2
Total	100%	100%	100%	100%	100%	100%	100%
Expected income change							
Increase	87%	74%	88%	90%	81%	67%	53%
A lot	17	12	13	28	26	22	14
A little	70	62	75	62	55	45	39
No change	5	13	9	5	9	22	24
Decrease	1	3	*	1	2	2	8
Don't know, not ascertained	7	10	3	4	8	9	15
Total	100%	100%	100%	100%	100%	100%	100%

* Less than 0.5 percent
a Includes those who were transferred within the same company.
b Not applicable.

	Transferred		Changed employer	
	Machine change	No machine change	Machine change	No machine change
Proportion who have had a promotion in past 5 years	57%	63%	38%	32%
Proportion who have had an income increase in past 5 years	89	94	75	71
Proportion who expect an income increase	89	90	90	79
Number of cases	(70)	(160)	(292)	(466)

In the latter group, workers with a concurrent machine change are slightly better off than those not changing equipment. This is the same pattern that is evident for the labor force as a whole. However among those who have been transferred, a concurrent change in equipment is associated with a slightly less favorable position vis-a-vis those without a machine change. The number of transfer cases is very small. Still, it is plausible that transfers occasioned by equipment change should involve upgrading less often than other transfers.

The data in Table 5-6 disregard the size of income increases. In the multivariate analysis which follows we shall make an additional distinction between larger and smaller income increases.

The foregoing paragraphs indicate the largely positive direct impact of machine change on income and careers. A further indication of this same fact is evident in people's expectations (lower part of Table 5-6). A very high proportion of the labor force, some 76 percent, felt that their incomes would rise in the next few years. Those persons with machine change were more heavily represented in this group than those without. Thus 87 percent of those with a machine change but no job change expected a rise in income (17 percent expected a large rise) while only 74 percent of those without either a job change or a machine change expected such a rise (12 percent expected a large rise). The same pattern is observed among the self-employed, although the self-employed as a whole seem to have more cautious expectations than other members of the labor force.

Despite the finding that those who changed jobs have poorer past income experience than those who did not, their expectations seem to be more buoyant. Apparently they are looking forward to the realization of increased income and position which the job change was supposed to accomplish. Job changers who also changed equipment had more optimistic expectations than those continuing to use the same equipment. Workers who left employers who were cutting back on production had somewhat depressed expectation; only 72 percent expected a rise in income versus 84 percent of other job changers. The job choice of this group again appears to have been adversely affected by their being no doubt subject to rather strong pressure to change jobs.

Much of the popular discussion about the unfavorable effects of automation centers upon the fact that machines can put people out of work. We saw in Section C of Chapter 4 that only a very small proportion (6 percent) of those who experienced a significant change in machine technology were unemployed or worked shorter hours at the time the new equipment was introduced. Yet, it is conceivable that in the longer run the increased capacity of the new machines would lead to additional layoffs or would make for less steady employment. To test this proposition, we shall now compare the unemployment during the past 5 years of people who did and those who did not experience changes in machine technology during the same 5 years.

For workers who remained with the same employer the employment related consequences of change in machinery are small. Table 5-7 indicates that if there are any differences in this group between those with machine changes and others, the machine changers are favored. Among those who had a machine change 8 percent were unemployed during the past 5 years, and among those who did not 11 percent were unemployed. The difference in the

TABLE 5-7

DISTRIBUTION OF WORKERS BY UNEMPLOYMENT AND MACHINE CHANGE

	Same job for past 5 years		Different job[a]		
	Machine change	No machine change	Machine change		No machine change
			Left old job because of machine change	Left old job for other reasons	
Number of periods of unemployment - past 5 years					
None	89%	88%	72%	55%	66%
One	4	6	22	21	16
Two	2	2	*	10	7
Three or more	2	3	3	11	8
Not ascertained	3	1	3	3	3
Total	100%	100%	100%	100%	100%

*Less than 0.5 percent.
[a]Includes those who were transferred within the same company.

frequency of repetitious unemployment between the two groups is insignif-icant; but prolonged unemployment seems to occur more often in technolog-ically static work situations.

Table 5-7 further indicates that there is a stong association between job change and unemployment. In some cases unemployment may force a worker to change jobs; in other cases a voluntary job change may entail some inter-im unemployment. About 35 percent of job changers were unemployed during the 5-year period studied, compared with only 9 percent of those with a single job. Repetitive unemployment affected 15 percent in the former group and only 5 percent in the latter. Thus the question becomes relevant again: To what extent is job change due to machine change? It was found in Chapter 3 that only 1.2 percent of our sample of the labor force said they changed jobs *because of* machine change, and most of these workers were not unemployed between jobs; two-thirds of them were transferred to new jobs within the same company, which frequently meant upgrading in terms of position and income. Indeed workers who changed jobs *because of* a change in equipment had a lower incidence of unemployment (25 percent) than the remaining group of job changers. They also show a very low frequency of repetitive unemployment, indicating that the unemployment occasioned by the job change was seldom followed by other layoffs. Still, this group *did* have more unemployment than people who could stay on the same job. More-over, we argued in Chapter 3 that the frequency with which workers them-selves attribute job change to machine change probably should be adjusted up-ward. For in a round-about way machine change may be responsible for un-acceptable working conditions more often than employees recognize; and these working conditions may lead to seemingly voluntary job changes. According to the evidence presented in Chapter 3, the needed upward adjust-ment to allow for "unrecognized" job change due to machine change is not very large. If this proposition is correct, it follows that unemployment due to machine change was of very moderate proportions during the period under study.

This conclusion is reinforced by the reasons which the unemployed themselves gave for their most recent period of unemployment. A significant proportion of workers stated that they were laid off because their former employer was cutting back production. These workers were found to have had a relatively low frequency of equipment change within the past 5 years. Some of these workers may have experienced unemployment because their employer failed to modernize his equipment rather than because he introduced new equipment.[6] We saw earlier that expanding firms are most likely to make

---

[6]This is one of the *indirect* consequences of machine change which this study can-not measure.

technological improvements. One may speculate further that the added productivity and skills acquired by an employee with new equipment are likely to
protect him from layoff when the company is having bad times. The newest
equipment often is the last to be shut down. By contrast, unemployment
blamed on recession, appeared equally often among all the groups in Table 5-7.
Workers with seasonal unemployment were highly represented in the group
with both a job and machine change.

Unemployment among job changers was high among those who had a
concurrent change in equipment—42 percent—in comparison to 31 percent for
those who did not. There was also more repetitious unemployment in the
former group. Yet in looking at unemployment we should realize that causation may run from unemployment to machine change as well as in the
opposite direction. Prolonged or repetitious unemployment may persuade a
person to accept a job in a different occupation, using different equipment.
The more frequent are job changes the more likely is a change in equipment.
Considering everything, for the years 1962 to 1967 very little unemployment
can be attributed *directly* to machine change. The indirect impact of machine
change on unemployment may be greater, but can only be guessed.

When asked about the "steadiness" of their new work, people who had
experienced advances in machine technology generally felt that their prospects
were improved. As seen in Table 5-8, many more workers with a machine
change felt that this change had increased the steadiness of their job and
raised their chances for advancement than believed the opposite. (Many jobs
are not subject to unemployment in the first place and therefore cannot
become more steady.) Workers without a job change but with an equipment
change felt in 21 percent of cases that their job was more steady, and in 23
percent of cases that it offered more opportunity for advancement after the
machine change. Job changers made such favorable evaluations of their new
job still more often. Forty-nine percent of workers changing both jobs and
machinery reported more steadiness, and 61 percent reported a greater chance
for advancement; but we must recall that obtaining more steady work and
bettering one's chances for advancement are among the primary reasons for
transfers and job shifts. A concurrent machine change is only incidental to the
search for greater steadiness and chance for advancement. Yet workers who
did change jobs *and* equipment seem more optimistic toward the future than
workers who changed jobs only. Workers who changed jobs *because of* an
equipment change reported somewhat fewer improvements in steadiness and
chances for advancement than job changers in general.

Negative reactions were very rare among those who had experienced an
equipment change without job change. Such reactions did occur most
frequently among those who changed both jobs and equipment, perhaps due
to the job change and not to the incidental equipment change. Those persons

TABLE 5-8

WORKERS' PERCEPTIONS OF STEADINESS AND ADVANCEMENT
IN RELATION TO MACHINE CHANGE

	Same job for past 5 years	Different job[a]		No machine change
	Machine change	Machine change		
		Left old job because of machine change	Left old job for other reasons	
Steadiness				
More	21%	26%	49%	31%
Same	69	52	36	51
Less	4	19	13	9
Not ascertained	6	3	2	9
Total	100%	100%	100%	100%
Chances for advancement				
More	23%	42%	61%	41%
Same	67	39	20	40
Less	3	16	17	10
Not ascertained	7	3	2	9
Total	100%	100%	100%	100%

[a]Includes those who were transferred within the same company.

who had changed jobs *because of* a machine change also made unfavorable evaluations in relatively large numbers. Perceived steadiness and chances for advancement then, are two of the very few variables (another being the length of unemployment) in which workers who have been forced to change jobs or transfer *because of* a change in machinery are somewhat worse off than other job changers.

The foregoing pages have explored the relation between machine change and the economic progress of the workers who experienced it. On the whole machine change appeared as a positive force, leading to rising incomes and promotions. Moreover, machine change does not have the strong deleterious effects on employment which are sometimes charged against it. The evidence presented in this chapter together with the evidence presented in Chapter 4 leads to the conclusion that very little *direct* unemployment of serious length was created by changes in machine technology during the mid-1960's. The group with whom one may be particularly concerned, workers who changed jobs *because of* a machine change, seemed to suffer no particularly serious hardships when compared to workers who changed jobs for other reasons. Of course the mere fact that these people were often forced to change jobs imposes some burden upon them, as workers who change jobs for whatever reason are as a group worse off than others in terms of past income change and unemployment. It is important then to note that the available evidence indicates that this group was not large.

If the direct economic results of machine change are predominantly favorable, there still is a significant minority of cases with unfavorable experiences. Although the percentages involved are very small, the number of workers is large in absolute terms.

C. Multivariate Analysis

We have explored the relationship of automation level of equipment to some measures of economic well-being. We have also compared the economic status of people directly affected by machine change with the economic status of people not so affected. Up to this point no allowance has been made for the fact that these groups differ in other respects as well, particularly in education. In order to discover how education and other variables affect the analysis we shall subject the data to Multiple Classification Analysis, a technique which is theoretically able to separate the influence which each of a given set of variables has on one dependent variable. For example, MCA should be capable of separating out the relation of automation level to income, allowing for the separate effect of education.

Conventional multiple regression analysis computes for each independent or explanatory variable one coefficient indicating its statistical influence on the dependent variable after the effects of other variables are taken into account. MCA, which is an extension of "dummy variable" multiple regression analysis, obtains such a coefficient for every *subclass* of every independent variable.[7] For every one of these subclasses MCA computes two coefficients. In each equation the mean value of the dependent variable is calculated for the sample as a whole. The first coefficient for each subclass is the amount by which the mean of the dependent variable calculated for the observations in that *subclass* deviates from the grand mean for the whole sample under analysis. This coefficient is given in the tables as the unadjusted deviation from the mean, and indicates to what extent the workers in a given subclass are below or above average with respect to the dependent variable. As in a two-way table, the unadjusted deviations show the simple relation between the

---

[7]See Daniel Suits, "The Use of Dummy Variables in Regression Equations," *Journal of the American Statistical Association,* December 1957, pp. 548-551.

dependent and the independent variables. The second coefficient of each sub-class is called the adjusted deviation from the mean and represents the deviation of each class mean from the grand mean, *after* the effects of all other variables in the analysis have been accounted for. In other words, the adjusted deviation indicates whether a given subclass is above or below average in some respect *after* allowing for other variables which may influence the class' behavior.

If we have information to classify every worker on the dependent and explanatory variables, then MCA enables us to predict (with a minimum sum of squares of errors), for instance, the probability that a given worker will report a greater opportunity for advancement now than 5 years ago. This probability is found by starting with the mean probability for the whole group, and adding or subtracting adjusted deviations according to the subclasses of the explanatory variables into which the worker falls. This is possible since additivity is assumed by MCA.

In the illustration below the dependent variable is coded 1 for those who perceive a greater chance for advancement than 5 years ago, and zero for all others. The mean probability for the whole sample is .484; that is 48.4 percent of the workers perceived increased chance for advancement. The adjusted deviations may be interpreted with reference to a hypothetical worker as follows:

Mean for the Entire Group: <u>48.4%</u>

<u>Adjustments</u>

1. Worker is under age 35	+ 4.6	
2. Worker is Negro	+ 2.2	
3. Worker had 8-11 grades of school		- 4.1
4. Works for multiplant firm with 50 to 499 employees in his branch	+ 4.6	
5. Blue-collar worker		- 7.8
6. Worker had a job change and an equipment change during the last 5 years	<u>+14.1</u>	<u> </u>
	+25.5	-11.9

Expected probability that a worker with the characteristics listed above will report greater chance for advancement: 48.4 + 25.5 - 11.9 = 62.0%.

In a similar way, the dollar value of income level or change may be used as the dependent variable. In this case, the mean as well as the positive and negative deviations would represent dollar amounts.

Finally, a simplified numerical scale may be used as the dependent variable. In Table 5-9 below the dependent variable, past income change, is scaled as follows:

0 Current income is same or less than 5 years ago

1 Current income is 101-114 percent of income 5 years ago

2 Current income is 115-139 percent of income 5 years ago

3 Current income is 140 percent or more of income 5 years ago

Deviations are merely numerical adjustments to the overall mean of 1.5 (which implies an increase in the range from 15 to 39 percent).

When one comes to evaluating the importance of various explanatory factors or their level of significance, it is usually a whole *set* of subclasses which is of interest, not just a single one. That is, the usual question is not how important being a high school graduate is, but how important education is. Therefore a beta coefficient is computed for each independent variable as a whole. The beta coefficients computed by the MCA program are analogous to the partial beta coefficients of regression analysis. They indicate the relative importance of each independent variable in explaining the dependent variable. For example, in Table 5-9 age emerges as the most influential variable (of those included in the equation) in regard to past income increases. Size of the firm for which a man works is next in importance, and occupation is a close third. Secondly for each equation the multiple R^2 is given. This statistic tells how effective all the independent variables jointly are in explaining movements in the dependent variable. For a further explanation of the MCA technique the reader is referred to Appendix IV.

The first dependent variable to be analyzed is income change over the past 5 years. We postulate that the factors making for income increases may not be exactly the reverse of the factors making for income decreases. Hence the income change variable was scaled in two different ways. For the analysis of income increases, (as shown above) workers with a decrease or unchanged income were coded 0, increases up to 15 percent were coded 1, increases from 15 to 39 percent were coded 2 and still larger increases were assigned a value of 3. The mean scale value for the entire sample of 2,662 cases is 1.5. For the analysis of income decreases, workers with a decrease were coded 1; and all those with an increase or no change were coded 0. The mean scale value for the sample is 0.5.

In order to analyze the determinants of income increases three MCA equations were fitted. The first related income increases to five socio-economic explanatory variables: age, size of firm, occupation, formal education and race. In addition to these variables, the second equation included job change, transfer, and equipment change; and the third equation included, in addition to the socio-economic variables, automation level of operated and indirectly used equipment.

The results of this three-step analysis are shown in Table 5-9. The larger the positive deviation of any class from the grand mean, the larger or more frequent are the income increases in that class. Age, size of firm, and occupation show a relatively strong relation to increased income. The incidence of income gains falls with advancing age. Income increases are more frequent or larger for workers in multiplant firms than those in single plant firms. They are least frequent among workers in small single plant firms. The incidence of income increases also is higher for professional and technical workers, the self-employed and craftsmen and foremen than for other occupational groups. The "adjusted" coefficients show a rather weak relation of education to the incidence of income increases, apparently because of the interrelation with occupation and age. Race is of negligible importance as a determinant of income increases after such factors as education and occupation have been taken into account; its beta coefficient is only .03.

In equations two and three of Table 5-9 the independent variables automation level and machine change were added to the basic equation. Since the relative effects of the socio-economic variables were virtually identical in the three equations, the deviations and beta coefficients associated with the socio-economic variables have not been repeated for equations two and three.

We hypothesized earlier in this chapter that the relation between the automation level of operated equipment and income change might appear much weaker after other variables, particularly education, were held constant. This is indeed the case. Automation level of operated equipment enters with a beta of only .06 which is low when compared to the beta of .23 for age in the same equation. The automation level of equipment used indirectly also is relatively insignificant. Equally important, the introduction of automation level does not raise the explanatory power of the equation significantly; the multiple R^2 are virtually identical—12 and 13%. Whereas the unadjusted deviations show a clear positive relation between income change and automation level the adjusted deviations are much smaller. It remains true, however, that those who operate equipment under numerical, tape, or computer control seem to obtain more and larger income increases than other groups of workers. It also remains true that those who have indirect contact with any kind of equipment have made more progress—in terms of income—than those who have no indirect contact with equipment at all.

Table 5-6 indicated that machine change raises the chance of income increases for most groups of workers. However when other variables are held constant, machine change falls into the background along with automation level as an influence on income change. Kind of work change carries a beta coefficient of only .07. Although machine change is not strongly related to

TABLE 5-9 (Sheet 1 of 2)

THE RELATION OF SOCIO-ECONOMIC VARIABLES,
MACHINE USE AND MACHINE CHANGE TO THE INCIDENCE OF PAST INCOME INCREASES

Mean of income increases: $\underline{1.52}$[a]

	Number of cases	Deviations from mean		Beta coefficient
		Unadjusted	Adjusted	
	Equation 1: Socio-economic variables only)			
Age				
Under age 35	663	+.35	+.34	
35-54	1,115	-.04	-.05	
Age 55 or older	430	-.45	-.40	
				.23
Size of firm's employment				
Single plant firm				
Under 50 employees	612	-.24	-.22	
50 or more	231	+.04	+.06	
Multiplant firm[b]				
Under 50 employees	350	+.18	+.16	
50-499	430	+.17	+.13	
500 or more	480	+.13	+.10	
				.16
Occupation				
Professional, administrative, and self-employed workers	638	+.19	+.19	
Clerical and sales workers	405	+.05	-.05	
Craftsmen, foremen	386	+.13	+.12	
Other blue-collar workers	773	-.25	-.19	
				.14
Formal education				
0-7 grades	185	-.47	-.16	
8-11 grades	641	-.16	-.07	
High school degree	695	+.05	*	
College[c]	663	+.24	+.08	
				.06
Race				
White	1,976	+.02	*	
Other	189	-.19	-.06	
				.03

Multiple R^2 = 12%

(See sheet 2 of this table for definition of above footnotes.)

TABLE 5-9 (Sheet 2 of 2)

THE RELATION OF SOCIO-ECONOMIC VARIABLES,
MACHINE USE AND MACHINE CHANGE TO THE INCIDENCE OF PAST INCOME INCREASES

Mean of income increases: $\underline{1.52}^{a}$

	Number of cases	Deviations from mean		Beta coefficient
		Unadjusted	Adjusted	
Automation level of operated equipment	(Equation 2: Automation level and socio-economic variables)			
Numerical, tape, computer, or other logical control	93	+.54	+.23	
Fixed mechanical control	832	+.006	-.04	
Powered multi-system, manual main control	351	-.14	+.05	
Manual control, operator powered or powered single-system	293	-.10	-.04	
No equipment	629	+.02	*	
				.06
Automation level of equipment used indirectly				
Numerical, tape, computer, or other logical control	148	+.41	+.14	
Fixed mechanical control	347	+.19	+.12	
Powered multi-system, manual main control	139	+.14	+.11	
Manual control, operator powered or powered single-system	36	+.20	+.21	
No equipment	1,523	-.11	-.06	
				.08

Multiple R^2 = 13%

	Number of cases	Deviations from mean		Beta coefficient
Kind of work change	(Equation 3: Machine change and socio-economic variables)			
Machine change only[d]	209	+.17	+.15	
No job or machine change[d]	1,188	-.13	-.03	
Change in employer and machine change	227	+.10	-.08	
Transfer and machine change	62	+.33	+.21	
Change in employer only	382	+.04	-.04	
Transfer only	147	+.38	+.17	
				.07

Multiple R^2 = 12%

---

*Less than .01.

[a]The dependent variable gave present income as a percent of income 5 years ago. It was coded: 0. 100 percent and under; 1. 101-114 percent; 2. 115-139 percent; 3. 140 percent or more. Positive deviations, therefore, imply larger than average increases in income.

[b]Number of employees in the respondent's plant or office.

[c]The group includes all those with college experience of at least one year.

[d]Includes the self-employed.

income change, we should note that its influence appears consistently positive. For the large group who remained with the same employer (including those who were transferred) machine change appears to have been advantageous financially.

Having discussed the impact of machine technology on income increases, it is sufficient to note briefly its relation to income decreases. The equations used to examine decreases were identical to the equations used above except for the dependent variable. However, the same predictor variables contribute much less to the explanation of decreases in income than they do to the explanation of income increases. Automation level shows only a weak and irregular relation to income decrease once other variables have been accounted for. On the other hand, kind of work change turns out to be of some importance. Its beta coefficient is .19, higher than for any other variable, the next highest beta being .11 for size of firm. A positive deviation from the mean here signifies more income decreases, a negative deviation, fewer. The analysis indicates that income decreases hardly ever occur except in connection with a job change. After "adjustments," the chance that a job changer's income decreased over the past 5 years was 12 in 100 for those who changed jobs only and 16 in 100 for those who changed jobs *and* equipment. The incidence of income decreases is not high, considering that job changers are the group among whom unemployment was concentrated (see the table below). Among people who stayed with the same employer over the past 5 years, the chance of an income decrease is 3 in a 100 if they had no machine change and 1 in 100 if there was some machine change.

THE RELATION OF MACHINE CHANGE TO INCOME DECREASES[a]
(After the basic socio-economic variables have been accounted for)

Mean of income decreases: <u>.06</u>

Kind of work change	Deviations from mean		Beta coefficient
	Unadjusted	Adjusted	
Machine change only[b]	-.05	-.05	
No job or machine change[b]	-.02	-.03	
Change in employer and machine change	+.07	+.10	
Transfer and machine change	-.01	*	
Change in employer only	+.05	+.06	
Transfer only	-.04	*	
		Multiple R^2 = 6%	.19

*Less than .01.

[a]Age, size of firm, occupation, education, and race are the other independent variables included in the analysis.

[b]Includes the self-employed.

A further dependent variable analyzed was one measuring income expectations. Earlier we pointed out the advantage of expected income change over past income change, in the sense that it more assuredly reflects the consequences of machine change, or at least the consequences of machine change as perceived by the workers. The income expectations variable was scaled as follows: 0 for expectations of an unchanged or declining income, 1 for "up a little," and 2 for "up a lot." The mean for the sample is .96. Automation *level* has only a very minor influence on the expectation of income increases. The beta coefficients are .04 and .05 respectively for automation level of equipment operated by the respondent and equipment with which he has indirect contact. Still, the weak relationship which exists shows optimism associated with rising automation level; the least optimistic workers are those who do not operate equipment and those who do not have indirect contact with equipment.

Kind of work change experienced is of some importance as a determinant of income expectations; the beta coefficient for this variable is .12. Table 5-6 indicated that, among people who had only one employer, those with a machinery change generally expected more and larger income increases than those without such a change. This relationship is preserved, though weakened, after socio-economic variables have been accounted for. It also remains true that job changers combine a relatively unfavorable past income history with relatively optimistic expectations.

Another variable which is related to expected income change is the worker's expectation of future changes in equipment in the company where he works. In order to measure people's perceptions of the technological progressiveness of their employer, they were asked: "Looking now into the future, do you expect that during the next few years the company you work for will modernize its machinery and equipment?" Workers answering "yes" or "perhaps" to this question reported significantly higher income expectations than those answering "no," as the data following indicate. These workers may have felt that the expected machine change would be beneficial to their incomes. A somewhat different interpretation is also possible. The modernization of equipment is the hallmark of a generally progressive or growing firm, and such firms are more likely to give income increases, whether they are equipment-related or not.

The relatively low R^2's obtained may surprise readers accustomed to aggregate time series or grouped data. At the individual level an immense variety of influences affect people's experiences. Many of these such as health, personality, and psychological characteristics have constant distributions over time or in large groups. R^2's well below 20 percent in individual cross-sectional analysis are common between the very variables which show very high R^2's in time series analysis or when grouped data are used.[8]

EXPECTED MACHINE CHANGE IN COMPANY
RELATED TO EXPECTED INCOME INCREASES**

(After allowing for the effects of
socio-economic variable)

Mean of expected increases in incomes: .96[a]

Expected machine change	Number of cases	Deviation from mean		Beta coefficient
		Unadjusted	Adjusted	
No change expected	1,038	-.18	-.12	
Change is possible	316	-.08	-.07	
Change is probable or certain	1,227	+.17	+.12	
				.17
Multiple R^2 = 19%				

**Age, size of firm, occupation, education, and race are the other independent variables included in the analysis.

[a]Those expecting no income increase were coded zero, those expecting a small increase were coded one and those expecting a large increase were coded two. Thus positive deviations from the mean are associated with higher income expectations.

Under these circumstances we can say with considerable confidence that machine change and the use of automated equipment do *not* have an adverse impact on the income trend of the workers concerned. We must be more cautious in accepting evidence that there is a small favorable income effect associated with equipment change and the use of technologically advanced equipment.

The final variable subjected to multivariate analysis is unemployment experience. This part of the analysis omits the self-employed, for whom unemployment is meaningless, as well as professional and administrative people and all college graduates, for whom unemployment is rare. Workers who had no unemployment experience in the past 5 years were coded 0; workers with one short spell of unemployment (one month or less) were coded 1; and workers with several short spells or one longer spell (over one month) were coded 2. The mean for the groups analyzed is .48. A positive deviation from the mean implies more serious unemployment. The results are shown in Table 5-10. The association of unemployment with low levels of education, youth, being non-white, or being a semi-skilled or unskilled blue collar worker is well known. The relatively high incidence of unemployment among employees of small firms is more interesting. Kind of work change appears to be by far the

---

[8]See for example Richard F. Kosobud and James N. Morgan, *Consumer Behavior of Individual Families Over Two and Three Years,* Survey Research Center, University of Michigan, 1964.

TABLE 5-10

THE RELATION OF SOCIO-ECONOMIC VARIABLES
AND MACHINE CHANGE TO THE SERIOUSNESS OF PAST UNEMPLOYMENT

Mean of unemployment experiences: $.48^a$

Age	Number of cases[b]	Deviations from mean		Beta coefficient
		Unadjusted	Adjusted	
Under age 35	664	+.13	+.06	
35-54	779	-.09	-.04	
Age 55 or older	290	-.06	-.02	
				.06
Race				
White	1,501	-.04	-.02	
Other	189	+.22	+.10	
				.06
Formal education				
0-7 grades	177	+.12	+.08	
8-11 grades	637	+.06	+.05	
High school degree	635	-.07	-.05	
Some college, less than B.A.	257	-.08	-.06	
				.07
Size of firm's employment				
Single plant firm				
Under 50 employees	400	+.11	+.08	
50 or more	202	+.18	+.14	
Multiplant firm[c]				
Under 50 employees	288	+.02	+.01	
50-499	398	-.05	-.01	
500 or more	407	-.18	-.12	
				.11
Occupation				
Clerical and sales workers	488	-.15	-.09	
Craftsmen, foremen	394	-.08	-.06	
Other blue-collar workers	841	+.13	+.09	
				.11
Kind of work change				
Machine change only	115	-.23	-.15	
No job or machine change	878	-.20	-.20	
Change in employer and machine change	241	+.44	+.40	
Transfer and machine change	55	-.12	-.02	
Change in employer only	342	+.32	+.29	
Transfer only	108	-.05	*	
				.30

Multiple R^2 = 14%

---

*
Less than .01.

[a]Workers with no unemployment in the past 5 years were coded 0; workers with
one short spell of unemployment (one month or less) were coded 1; and workers
with several short spells or one long spell (over one month) were coded 2.
Positive deviations from the mean therefore indicate more serious unemployment.

[b]The self-employed, those in the professional and business occupations, and all
college graduates were excluded from this analysis.

[c]Number of workers in the respondent's plant or office.

most important determinant of unemployment, with a beta coefficient of .30. The main distinction is between job changers and all others, rather than between machine changers and all others. Among workers with only one employer, unemployment was marginally higher if a worker had an equipment change than if he did not. Chapter 4 showed that machine changers who do not have to change jobs occasionally experience some unemployment during the transition to new equipment. People who had a job change had relatively high rates of unemployment. This is not surprising since unemployment may cause a job change as well as being a consequence of it; causation runs both ways. Those workers who changed jobs *and* equipment had more unemployment than those who changed jobs only; but it is quite possible that unemployment makes a person more disposed to take a job which differs considerably from the previous one. Thus the equipment change may be a consequence of the unemployment. We have no clear evidence that it contributes to the unemployment of job changers.

The adjusted relations of automation level to unemployment are weak and irregular (beta coefficients of .07 and .08 respectively for equipment operated by the worker and equipment used indirectly). The earlier indication (Table 5-4) that unemployment is somewhat less likely at high automation levels than at the lower ones reflects largely the association of education with sophisticated equipment.

The conclusions drawn in Section B are not fundamentally altered by the multivariate analysis. To the extent that technological change forces or induces people to change jobs, it contributes to unemployment. However, we have gone to some length earlier to show that the proportion of people whose job change can be attributed to machine change is small, even if some allowance is made for understatement by respondents. We have seen above that, despite the higher incidence of unemployment among job changers, past income increases were only slightly less prevalent (after "adjustment") than among those without job change; income decreases were somewhat more frequent but still uncommon. The data further indicate that job changers are particularly optimistic about the steadiness of their new job, the chances for advancement, and their income prospects.

Among people who had no job change, the chance of unemployment seems to be enhanced very slightly by equipment change. Moreover, this weak unfavorable relationship exists only in the abstract, i.e. when level of education is held constant and groups in the labor force which are not susceptible to unemployment are disregarded. The point is that the well-educated as well as technical, administrative and professional people are used extensively to work with the more mechanized and automated categories of equipment. And when their employment record is taken into account, machine change appears to be associated with less rather than more unemployment. To be sure, the "cause" of the better employment experience seems to be the professional and educational status of the people who work with the newer machines, rather than the machine use or machine change itself.

D. Job and Geographic Mobility in Relation To Machine Change

The study of job and geographic mobility has many facets which are unrelated to technological change. We may, however, explore briefly the relation between equipment change and mobility in this section.

The effects which automation level and machine change could conceivably have on job and geographic mobility do not point clearly in one direction. The use of highly automated equipment, or a change from one piece of equipment to another, may involve the acquisition of new skills. These skills would most likely be in demand and put the worker in a more favorable labor force position, broadening his range of job opportunities and making him more mobile. He may be more or less satisfied with his job after the equipment change, and such attitudes would also affect his moving plans. In some cases a change to new equipment may make a worker's job less secure or steady and his skills obsolete. The worker's weakened labor market position would then limit his job opportunities; yet he might increase his effort to find other work. A further consequence of machine change might arise when a worker shifts to highly specialized equipment. The use of such equipment might develop very specific skills and experiences useful only to a few employers. The worker might then have to choose between remaining with his present employer or making a geographic move.

The major problem in determining any relation between past equipment change and past job mobility is that causation runs both ways. Not only can equipment change affect job mobility, as mentioned in the last paragraph, but a job change often involves a change in the equipment with which a person works. Relations between past equipment change or current technological level of equipment and *expected* mobility can be examined without encountering the same difficulties.

These relations show no consistent tendency for fairly definite plans to change jobs or to move to rise (or fall) with automation level of equipment. However, the conditional answer "possibly" is more frequent among people who work with logically controlled equipment than among others as regards both kinds of moves. It would appear that this group is particularly receptive to job offers or other opportunities which might present themselves, even if it meant moving to another county or state. We know that geographic mobility is enhanced by education; and logically controlled equipment is likewise associated with education. Thus the stimulus to geographic mobility comes at least partly from education.

Job changers are considerably more likely than those who have remained with the same employer during the past 5 years to have plans to change jobs or to make a geographic move. This finding reflects the well-known fact that planned mobility is associated with past mobility. There are no significant differences in planned mobility between those with and without an equipment change.

Plans or desires to change jobs in a sense imply dissatisfaction with the present job or at least a feeling that better alternatives are available. The finding that equipment change or the use of automated equipment do not of themselves give rise to plans to move or to change jobs is consistent with other evidence in this study that advances in machine technology seldom detract from the economic or non-economic rewards of the job.

—

Chapter 6

THE IMPACT OF CHANGES IN MACHINE
TECHNOLOGY ON PERCEIVED
JOB CHARACTERISTICS AND
JOB SATISFACTION

". . . the understandings of the greater part of men are necessarily formed by their ordinary employments. The man whose life is spent in performing a few simple operations, of which the effects too are, perhaps, always the same, or very nearly the same, has no occasion to exert his understanding, or to exercise his invention in finding out expedients for removing difficulties which never occur. He naturally loses, therefore, the habits of such exertion, and generally becomes as stupid and ignorant as it is possible for a human creature to become. The torpor of his mind renders him, not only incapable of relishing or bearing a part in any rational conversation, but of conceiving any generous, noble, or tender sentiment. . . . His dexterity at his own particular trade seems, in this manner to be acquired at the expense of his intellectual, social (and) martial virtues."[1]

---

*This chapter is largely the work of Judith Hybels.
[1] Adam Smith, *The Wealth of Nations,* New York, The Modern Library, 1937, pp. 734-735.

A long line of thought has followed in the tradition of Adam Smith, deploring the destructive impact of the machine on the human spirit.[2] In this age of affluence, more than ever, it is recognized that a job should meet certain non-economic needs of the work force—the need for achievement, for responsibility, for variety, for social contact and communication, for influence and control, and for physical comfort. Changes in such job characteristics are the subject of this chapter.

In Section A the data will be briefly described. Section B deals with changes in job satisfaction in relation to machinery and job change. We shall investigate whether some socio-economic groups derive more satisfaction from technological change than others. We shall also investigate how perceived changes in specific job characteristics affect job satisfaction. In Section C changes in job characteristics are linked with the kind of work changes which brought them about, particularly changes in machine technology. Throughout this analysis it appears that, in the view of the workers affected, changes in machine technology tend to make jobs more interesting and more demanding in a non-physical sense. It also appears that jobs which become more demanding tend to become more satisfying. We must suspect that this generalization does not hold in every case. Therefore in Section D we shall first look at economic conditions and socio-economic characteristics which are associated with this dominant pattern. And then we shall investigate socio-economic characteristics associated with the deviant pattern—a preference for jobs which have become less challenging in terms of skills, variety, responsibility, autonomy, and need to learn new things.

A. The Data Used

With a view toward measuring changes in job satisfaction, workers were asked:

"Taking everything into consideration, how satisfied are you with your job now as compared to 5 years ago? Are you more satisfied, about the same, or less?"

This question came after a detailed inquiry about changes in job content, income changes, attitudes toward automation and the new job, which pre-

---

[2]See for example the study by Chris Argyris, *Personality and Organization,* New York, Harper & Brothers, 1957, which describes how the task specialization, repetition, machine pacing, and minimal use of skills of assembly line work lead to employee apathy, dissatisfaction, and emphasis on material rewards at the expense of human ones. This kind of analysis focuses on extreme examples of machine-pacing which this study shows to be (proportionally) of little importance in a cross-section of the labor force.

sumably set a frame of reference for the job satisfaction question. The answers regarding change in job satisfaction will serve as the dependent variable throughout most of this chapter. In Chapter 7 a more comprehensive variable, intended to measure the worker's success in adjusting to technological change, is constructed and analyzed.

In addition to the question about overall job satisfaction, a battery of 16 attitudinal questions was asked in a different part of the interview to measure the perceived changes in a variety of job characteristics. These questions were addressed only to respondents who experienced an equipment and/or job change. Specifically, these people were asked to comment on the following dimensions of their job (the exact wording of the questions can be seen in questionnaire A, Section F of Appendix 1 of this volume):

Is the job more interesting or more monotonous?
Is it more or less steady?
Does it provide a greater or lesser chance for advancement?
Is the physical work increased or decreased?
Is the speed with which the respondent has to work increased or reduced?
Is more or less skill required of the respondent?
Is there more or less opportunity to learn new things?
Is there more or less need for planning, judgment, or initiative?
Has the respondent more or less influence in organizing the work?
Is the quality of the product more or less dependent on the respondent?[3]
Are errors more or less serious?
Is there more or less paper work?
Is the respondent more or less closely supervised?
Are the physical surroundings more or less pleasant?
Is there more or less chance to talk with fellow workers?
Is there more or less danger of personal injury?

The entire sequence of questions was posed in terms of *change* rather than absolute level for several reasons. First, we are interested in changes in various aspects of job satisfaction which are related to machine and/or job change. Changes can be identified by asking people to compare their former work with their work after getting the new machine or the new job. A man may consider his job to be demanding or interesting in an absolute sense *both* before and after a change, and yet he may feel that it has become more or less demanding and interesting as a result of the change. Secondly, the absolute level of interest, chances for advancement, or opportunity to talk are

---

[3]This question was asked of production workers only.

difficult to measure, particularly under circumstances which differ so greatly from worker to worker. Therefore interpersonal comparisons of such variables as the level of job satisfaction or perceived skill requirements are much less feasible than comparisons of, say, income level. Direction of change can, however, be elicited without reference to the absolute level of an attitude and can be compared between people and groups of people.

For purposes of analysis, people are split into three groups: (1) those who experienced a machine change only (including wage and salary workers and the self-employed); (2) those who experienced a job change or transfer and are using different machinery on the new job (a few of the job changes and transfers occurred *because of* machinery change; many more did not); (3) those who experienced a job change or transfer but no machinery change. In a sense the third group is used as a standard of comparison. That is, one may ask how changes in job satisfaction resulting from technological change compare with changes in job satisfaction associated with job shifts and transfers. It must be realized however that this is a rather rigorous standard. In a prosperous economy people often change jobs voluntarily with the hope of upgrading themselves financially, finding more congenial or challenging work, or escaping from an unsatisfactory work setting. Not surprisingly then, shortly after a job change, most workers feel that they have improved their situation. We wish to find to what extent this is also true of workers who have experienced a change in machine technology, largely non-discretionary on their part.

Before this question can be answered, we must recall that the three groups to be compared are not entirely alike with respect to such characteristics as age, income, and education (which also have a bearing on adjustment to technological and job change). Although the socio-economic differences are not large, we shall again make use of Multiple Classification Analysis in an attempt to discover the separate impact of job and equipment change on various aspects of job satisfaction.

One might suspect that people's evaluations of changes in various aspects of their job content would be correlated with each other because of what is often called a "halo" effect. If a worker likes his new equipment or work better than his former one, all of the responses may be biased toward the "better" side. He may have forgotten some particular advantages of his former job situation and judge the new one superior on almost all points because of one or two outstanding features. Conceivably the change in his income might color all other reactions to his new work. We shall return to this problem in the next section, so it suffices to say here that people generally reacted quite differently to various aspects of equipment and job change.

B. Change In Job Satisfaction and Some of Its Determinants

On the whole, job satisfaction is increased by changes in machine technology. While almost 60 percent of workers experiencing a change in machinery said that they were "more satisfied" with their work than they were 5 years ago, only 8 percent of these same workers felt that they were less satisfied. This compares to a proportion reporting an increase in job satisfaction of only 40 percent among workers without either a job or equipment change. Job change has a more positive impact on job satisfaction than does machine change. As noted earlier, many job changes are due to a voluntary search for better paying or more congenial work. It is not surprising that this search should very often be at least subjectively successful. It is significant however that workers who on the whole are passively associated with a change in equipment should more closely resemble job changers than they do workers who experienced no change of either kind. Job changes which are associated with equipment changes imply an even greater departure from the previous job content than job changes which involve no change in equipment. Not surprisingly, people in this category were least likely to report "no change" in job satisfaction (and indeed in all the measured components of job satisfaction) and most likely to report changes for the better as well as for the worse.

TABLE 6-1

CHANGE IN JOB SATISFACTION OVER THE PAST 5 YEARS
AMONG JOB AND MACHINE CHANGE GROUPS

Change in satisfaction	All	Same job for past 5 years[a]		Different job	
		Machine change	No machine change	Machine change	No machine change
More	51%	58%	40%	67%	64%
Same	34	34	45	16	20
Less	9	8	9	11	9
Not ascertained	6	*	6	6	7
Total	100%	100%	100%	100%	100%
Number of cases	(2,662)	(224)	(1,423)	(363)	(642)

[*] Less than 0.5 percent.

[a] Includes the self-employed.

The question asked was: "Taking everything into consideration, how satisfied are you with your job now as compared to 5 years ago--are you more satisfied, about the same, or less?"

Even people with no equipment change *and* no job change tend to view the progress of their jobs favorably. They expressed "more" job satisfaction than 5 years ago in 40 percent of cases. People have a well-known tendency to adjust aspirations to a level which is consonant with given possibilities. This consideration may not be very important here, since *changes* in satisfaction rather than the *state* of satisfaction are measured. Irrespective of any subjective bias, many people do advance financially over a period of years, improve their skills, get to know their co-workers better, and are given more responsibility. Especially for younger and well-educated workers, careers *do* follow an upward path in most occupations, and jobs should therefore become more satisfying.

In order to separate the various influences on job satisfaction a series of multivariate analyses was made using Multiple Classification Analysis. MCA has been described briefly in Chapter 5 and is explained at greater length in Appendix IV. A dependent variable was constructed with 1 signifying increased job satisfaction and 0 all other answers. A value of 60 percent for a particular group would signify that within that group 60 percent reported more job satisfaction or alternatively, that the probability of a person in that group being more satisfied was .60. The factors which are associated with job dissatisfaction need not be a mirror image of those which are associated with job satisfaction; in order to allow for this possibility a parallel multivariate analysis was carried out with a second dependent variable which distinguished between people who were less satisfied with their job (coded 1) and all others (coded 0). The two analyses are shown side-by-side in Table 6-2.

The analysis of increased job satisfaction includes only people who had an equipment change, sometimes with and sometimes without a job change or transfer. The analysis of job dissatisfaction includes in addition workers who had a job change but no equipment change. The reason for the enlargement of the sample was that the number of cases reporting less job satisfaction is rather small. In both parts of the analysis the kind of work change experienced is introduced as an independent variable. Such work changes include changes in machinery, changes in employers, transfers within the same company and combinations of these. The analysis begins by relating a basic set of economic and demographic predictors to job satisfaction. Later, variables measuring changes in perceived job characteristics are added one at a time to the basic set of predictors. This procedure enables us to study the impact of changes in particular job characteristics on job satisfaction, holding constant interrelated economic and demographic characteristics.

Table 6-2 shows the influence on job satisfaction and dissatisfaction of the basic set of predictor variables: age, occupation, education, income level, income change, and kind of work change experienced. The top row of the table indicates that among all people with an equipment change 63.5 percent

TABLE 6-2

RELATIONSHIP OF BASIC DEMOGRAPHIC AND ECONOMIC VARIABLES TO JOB SATISFACTION

		More satisfied[a]			Less satisfied[b]	
	Mean:	63.5%		Number	9.2%	
	Number	Deviations from mean		of	Deviations from mean	
	of cases	Unadjusted	Adjusted	cases	Unadjusted	Adjusted
Income change - past 5 years						
Decrease	34	-34.1%	-36.7%	79	+33.8%	+35.3%
No change	177	-5.4	-2.5	354	+2.1	+2.0
Small increase	132	*	+1.7	268	-4.3	-4.5
Large increase	155	+16.5	+14.7	326	-4.3	-4.3
Beta coefficient		.26			.34	
Kind of work change						
Machine change only	224	-6.0%	-8.4%	224	-1.6%	*
Change in employer and machine change	292	+5.6	+7.5	292	*	-1.3%
Transfer and machine change	71	-4.4	-4.2	71	+4.9	+6.5
Change in employer only	a	a	a	466	*	-1.1
Transfer only	a	a	a	176	+0.5	+2.6
Beta coefficient		.16			.07	
Occupation						
White collar	250	+3.7%	+2.9%	567	-2.7%	-2.4%
Blue collar	331	-1.9	-1.3	652	+2.2	+1.9
Beta coefficient		.11			.08	
Age						
Under age 25	107	+4.7%	+3.5%	214	-0.8%	+2.7%
25-34	179	+8.5	+5.2	367	-1.8	-1.3
35-44	144	-3.8	-4.1	295	-1.1	-1.4
45-54	95	-6.7	-3.4	214	+1.6	-0.9
Age 55 or older	61	-14.4	-7.3	133	+5.8	+3.5
Beta coefficient		.10			.07	
Formal education						
0-7 grades	35	-9.3%	*	75	+1.5%	-3.2%
8-11 grades	154	-1.2	+1.6%	310	+1.5	*
High school degree	201	-1.4	-3.3	415	+1.2	+1.9
College	190	+5.4	+3.1	410	-2.1	-0.6
Beta coefficient		.08			.07	
1966 income from worker's job						
Under $4,000	175	-4.7%	-4.2%	374	+0.7%	-1.2%
$4,000-5,999	131	*	+2.4	274	+2.1	*
$6,000-9,999	193	+1.7	-1.0	387	-1.4	*
$10,000 or more	86	+6.2	+7.4	182	-1.0	+2.0
Beta coefficient		.08			.04	
		R^2 = 7.7%			R^2 = 10.8%	

---

*Less than 0.5 percent.

[a]Includes only those workers who have experienced a machine change.

[b]Includes only those workers who have experienced a machine and/or job change.

were more satisfied after the change; and among all people with an equipment and/or job change 9.2 percent were less satisfied. Within the basic set of demographic and economic variables, income increases stand out as an important determinant of greater job satisfaction; income decreases are even more crucial in relation to job dissatisfaction. As expected, young people reported more frequently than older ones that their job is more satisfying than 5 years ago. Reports of less job satisfaction are uncommon in all groups; they occur with above-average frequency among the youngest and the oldest workers.[4] White collar occupations seem to be slightly more conducive to growing job satisfaction than blue collar occupations, while the relation of changing job satisfaction to income level and education is negligible. After accounting for the effects of other variables, kind of work change remains an important determinant of job satisfaction, as the relatively high beta coefficient indicates. Workers who had a job change with an equipment change were somewhat more satisfied than those who had an equipment change only. Workers with a machine change only appear in this analysis with a negative coefficient, i.e. as having a below average chance of greater job satisfaction. The reason for this negative showing is that this group is compared with job changers and transfers only. When people with no job *and* no equipment change are added to the analysis, this no-change group has a sizable negative coefficient, while the adjusted coefficients for all other groups deviate in a positive direction. Job changers show a larger positive deviation than equipment changers. As indicated in Chapter 5, job change is associated with optimistic expectations, and this may in part explain the association of job change with increased job satisfaction.

In addition to the basic set of predictors, the impact on job satisfaction of a number of other potentially relevant socio-economic variables was explored. These variables are presented in Table 6-3 in decreasing order of importance (beta coefficients), allowance again being made for the influence of the basic set of predictors. As before, the analysis is confined to people who experienced an equipment change with or without an accompanying job change. The data indicate that automation level of equipment has relatively little bearing on job satisfaction. The beta coefficient of .13 reflects primarily the comparatively low satisfaction of people who operate no equipment. Men are more likely to report increased job satisfaction than women. On the other hand, size of firm, race, absence of unemployment, union membership, and vocational education are of negligible importance in determining whether a person is more satisfied after a machinery change than before. Similarly, none

---

[4]People who were not working 5 years ago were excluded from this analysis; therefore very few of the youngest age group would have had a college education; many are high school graduates, but some are dropouts.

TABLE 6-3

IMPORTANCE OF RELATIONSHIP OF SOME ADDITIONAL DEMOGRAPHIC
AND ECONOMIC VARIABLES TO JOB SATISFACTION

	Beta coefficients	
	More satisfied[a]	Less satisfied[b]
Automation level - own equipment	.13	.04
Sex	.12	.07
Growth of firm	c	.11
Automation level - other equipment	.08	.07
Size of firm	.07	.09
Race	.07	.02
Unemployment	.06	.10
Union membership	.05	.004
Vocational education	.01	.02
Number of cases	(587)	(1,229)

[a]Includes only those workers who have experienced a machine change.
[b]Includes only those workers who have experienced a machine and/or job change.
[c]Beta coefficient of .10 resulting almost entirely from infrequent increases
in job satisfaction among a relatively large group who did not know the
direction of change in the work force in the place where they work.

of these variables, including sex, have an appreciable influence on satisfaction
for workers with a job change only (data not presented here).

Two variables—unemployment experience and past changes in the firm's
labor force—are of some importance as determinants of *de*creased job satisfaction. The experience of at least one long spell or several short spells of
unemployment is associated with reports that the job has become less satisfying after the machine and/or job change. Similarly, working in a place which
has a smaller work force than 5 years ago is also associated with reduced job
satisfaction. Whether the decline in the work force is due to technological
advance or a decline in the demand for the employer's products or services,
such a situation is bound to make for job insecurity. In the latter case the
workers may unfavorably compare their firm's stagnant or declining volume
of business with that of more successful competitor firms or industries. The
mean percentage of all job and/or machine changers who felt less job satisfaction is 9.2. The corresponding figure is 14.2 percent for those with at least
one long or several short spells of unemployment (after adjustment for interrelated demographic and economic factors, especially income change), and it is
15.2 percent for those working in a place with a declining labor force. It
should be noted that the positive counterparts of these two factors (having no

significant unemployment and working in a place with a constant or growing work force) make only a negligible contribution to greater job satisfaction.

A change in machine technology is frequently accompanied by changes in job content, work organization, job demands, and other job characteristics. We shall now see that these factors have a more important bearing on the worker's reaction to equipment change than do most of the socio-economic factors analyzed so far. The one exception is income change which retains its dominant role in relation to job satisfaction.

Table 6-4 relates both income change and changes in specific job characteristics to job satisfaction. In column 1 are given the simple correlation coefficients for the associations between change in job satisfaction on the one

TABLE 6-4

CORRELATION OF 16 JOB CHARACTERISTICS WITH JOB SATISFACTION AND INCOME CHANGE[a]
(Unadjusted data)

Change in:	Simple correlation coefficients	
	Change in job satisfaction	Income change
Income	.30	--
Opportunity to learn	.27	.08
Interest	.27	.10
Chance for advancement	.25	.10
Planning, judgment, initiative required	.20	.08
Skill required	.19	.14
Own influence on work	.15	.03
Steadiness	.12	.04
Seriousness of errors	.11	.09
Chance to talk	.10	.03
Pleasantness of physical surroundings	.08	.03
Influence on quality of production	.06	-.01
Closeness of supervision	.03	-.02
Amount of paper work	.03	.03
Danger of personal injury	.02	-.01
Speed required	-.01	.05
Physical effort required	-.05	-.02
Number of cases	(1,161)[b]	(1,161)[b]

[a] For the entire correlation matrix of these variables, see Table 6-11.

[b] Includes only those workers who have experienced a machine and/or job change; excludes cases which were not ascertained on several variables.

hand, and change in income and in each of the 16 measured changes in job characteristics on the other. The second column relates each of the 16 job variables to income change, again using the simple correlation coefficient as a measure of closeness of relationship. On this basis income change does indeed appear as the single most important determinant of job satisfaction, with a correlation coefficient of .30. However, three other changes in job characteristics seem to have an almost equal influence: whether the job provides more or less chance to learn new things (.27), whether the job is more or less interesting (.27), and whether the job provides more or less chance for advancement (.25). In each case the answer "more," is associated with more job satisfaction. Other fairly pronounced relations associate changes in job satisfaction positively with changes in the need for planning and judgment (.20) and with changes in required skill level (.19). The implication of column 2 which relates changes in perceived job characteristics to income change, is quite clear: None of the perceived changes in job characteristics are highly correlated with income change. The highest correlation is that between income change and required skill level (.14). It follows from Table 6-4 that income change itself does have an important bearing on job satisfaction, but that it does not cast a halo over other perceived job characteristics.

The most striking finding which emerges from Table 6-4 is that those changes which meet the worker's need for achievement and responsibility are the most crucial for increased job satisfaction. Aside from income change and interest, the job characteristics which are most highly correlated with job satisfaction are those which measure the perceived challenge of the new work: chances for advancement, the opportunity to learn, being able to plan and organize one's own work, and higher skill requirements.[5] Even having a job where errors are more serious is positively correlated with job satisfaction.

It has been said that job security is a particularly highly valued job attribute.[6] However, for the American labor force under present-day business conditions this generalization is not correct. The simple correlation between change in steadiness of work and change in job satisfaction is only .12. The explanation may be that steadiness is taken for granted by many people; a large majority of American workers have never been unemployed. The opportunity to talk, as with steadiness and pleasant surroundings, is only of secondary importance to the average worker because he is accustomed to satisfactory conditions. Jobs which have come to require more physical work or more speed are less satisfying, but the correlations are low. Indeed physical

---

[5]See Floyd Mann and L. Richard Hoffman, *Automation and the Worker,* New York, Henry Holt and Company, 1960, especially Chapter 4, for similar results from a case study.

[6]See for example, S. D. Anderman ed., *Trade Unions and Technological Change,* George Allen & Unwin Ltd., London, 1967, p. 149.

demands, speed, and danger of personal injury come close to being matters of indifference to the labor force as a whole, most probably because few jobs today are seriously objectionable on these counts.

Do these perceived job characteristics continue to make a large contribution to job satisfaction when income change and other basic socio-economic variables are held constant? To answer this question, perceived job characteristics were added one at a time to the basic set of predictors to determine their separate influence on job satisfaction. Table 6-5 shows clearly that for people with equipment change, even when the effects of demographic characteristics are allowed for, changes in factors such as interest, chance for advancement, opportunity to learn, and need for planning, judgment, and initiative remain important determinants of changes in job satisfaction. In fact, several of these job characteristics show a more pronounced influence on job satisfaction than do the personal characteristics investigated.

Among the job characteristics that diminish job satisfaction, the most decisive are a more monotonous job, less chance for advancement or learning, and less need for planning and judgment; these are the reverse of the factors

TABLE 6-5

IMPORTANCE OF RELATIONSHIP OF JOB CHARACTERISTICS TO JOB SATISFACTION[a]

	Beta coefficients	
Change in:	More satisfied[b]	Less satisfied[c]
Interest	.26	.31
Chance for advancement	.21	.20
Opportunity to learn	.20	.22
Planning, judgment, initiative required	.18	.18
Steadiness	.17	.08
Speed required	.15	.05
Skill required	.14	.16
Seriousness of errors	.11	.10
Physical effort required	.11	.03
Own influence on work	.11	.12
Number of cases	(587)	(1,227)

[a]Also included in each regression were: income change over the past 5 years, kind of work change in past 5 years, occupation, age, education, and 1966 income from worker's job.

[b]Includes only those workers who have experienced a machine change.

[c]Includes only those workers who have experienced a machine change and/or job change.

making for greater satisfaction. A reduced need for skill and smaller chance to influence one's own work are somewhat more important in relation to job dissatisfaction than are their positive counterparts in relation to job satisfaction. In this connection, it is interesting that a worker's having skills which he would like to use, but cannot use on the present job, also makes for dissatisfaction (beta coefficient .15), yet the absence of unused skills does not contribute appreciably to job satisfaction (beta coefficient .06). As indicated earlier increased speed and physical effort, as well as less steadiness, have comparatively little negative influence on job satisfaction.

One limitation of the analysis in this section has still to be mentioned. It was not possible within the scope of this study to look at changes in job content in a context of employee-management relations. Yet . . . "Experience shows that from a practical point of view the success of a sharp change of any kind which management introduces is as dependent on the climate of relationships at the time as upon the character of the change itself or even on the way in which it is administered."[7] Although the results presented here may to some extent be influenced by employee-management relations, it is unlikely that the overall climate of relationships should seriously bias our rather detailed inquiry into changes in job characteristics.

C. How Machine Change Alters Perceived Job Content and Job Characteristics

Having analyzed various job characteristics and their subjective meaning to the worker, we next want to ₍now how frequently and in what ways these job attributes are altered by changes in machine technology. In Table 6-6 three groups are compared: (1) those with an equipment change only, (2) those with a job change or transfer *and* a machinery change, and (3) those with a job change or transfer who continued to use the same kind of equipment. We find that technological change is positively associated with all but one of the changes in job characteristics which add to work satisfaction.[8] The one important exception is faster work pace. The reader may recall that there

---

[7]Charles R. Walker, *Toward the Automatic Factory: A Case Study of Men and Machines,* New Haven, Yale University Press, 1957, p. 155.

[8]In examining Table 6-6 it must be kept in mind that some favorable changes in job characteristics might have been reported even by workers without job or equipment change, just as these people reported on balance greater work satisfaction. In a study by Einar Hardin of automation in an insurance company, which used questions similar to those asked here, there was some tendency to report favorable changes in job characteristics, even in unaffected departments. See Einar Hardin, "The Reactions of Employees to Office Automation," in U. S. Department of Labor, Bureau of Labor Statistics, Bulletin No. 1287, *Impact of Automation,* 1960, pp. 101-108.

TABLE 6-6 (Sheet 1 of 2)

CHANGES IN JOB CHARACTERISTICS AMONG JOB AND MACHINE CHANGE GROUPS
(Unadjusted data)

Change in:	All[b]	Same job for past 5 years Machine change	Different job[c] Machine change	No machine change
Speed required				
More	32%	48%	36%	24%
Same	43	44	36	47
Less	20	5	26	21
Not ascertained[a]	5	3	2	8
Total	100%	100%	100%	100%
Physical effort required				
More	25%	15%	36%	21%
Same	38	34	27	46
Less	32	47	35	25
Not ascertained[a]	5	4	2	8
Total	100%	100%	100%	100%
Interest				
More	64%	60%	75%	59%
Same	21	31	10	24
Less	10	5	14	10
Not ascertained[a]	5	4	1	7
Total	100%	100%	100%	100%
Skill required				
More	49%	53%	59%	41%
Same	33	36	20	40
Less	12	7	18	11
Not ascertained[a]	6	4	3	8
Total	100%	100%	100%	100%
Opportunity to learn				
More	57%	49%	71%	51%
Same	28	42	14	32
Less	9	3	14	9
Not ascertained[a]	6	6	1	8
Total	100%	100%	100%	100%
Planning, judgment, initiative required				
More	54%	48%	62%	51%
Same	32	43	22	35
Less	8	5	14	6
Not ascertained[a]	6	4	2	8
Total	100%	100%	100%	100%

[a] Includes not working 5 years ago.

[b] Includes only those workers who have experienced a machine change or a job change or both.

[c] Includes those who were transferred within the same company.

TABLE 6-6 (Sheet 2 of 2)

CHANGES IN JOB CHARACTERISTICS AMONG JOB AND MACHINE CHANGE GROUPS
(Unadjusted data)

| | | Same job for past 5 years | Different job[c] | |
Change in:	All[b]	Machine change	Machine change	No machine change
Own influence on work				
More	46%	34%	53%	44%
Same	37	54	25	39
Less	12	7	20	9
Not ascertained[a]	5	5	2	8
Total	100%	100%	100%	100%
Seriousness of errors				
More	38%	31%	52%	32%
Same	38	53	21	44
Less	18	11	25	15
Not ascertained[a]	6	5	2	9
Total	100%	100%	100%	100%
Influence on quality of production (Production workers only)				
More	45%	29%	63%	35%
Same	26	37	24	25
Less	8	10	8	6
Not ascertained[a]	21	24	5	34
Total	100%	100%	100%	100%
Pleasantness of physical surroundings				
More	37%	20%	49%	36%
Same	41	66	28	40
Less	17	10	21	16
Not ascertained[a]	5	4	2	8
Total	100%	100%	100%	100%
Danger of personal injury				
More	22%	15%	37%	15%
Same	48	56	32	56
Less	25	25	30	22
Not ascertained[a]	5	4	1	7
Total	100%	100%	100%	100%
Chance to talk				
More	34%	20%	47%	32%
Same	43	64	28	45
Less	18	12	24	15
Not ascertained[a]	5	4	1	8
Total	100%	100%	100%	100%
Number of cases	(1,175)	(170)	(363)	(642)

(For definition of above footnotes see page 1 of this table.)

is a weak negative correlation between increased work pace and work satisfaction. Having to work faster often means having to pay closer attention and this adds to the strain and tension of work life. Table 6-6 shows that nearly one-half of those who experienced a change in machine technology said that the speed with which they have to work was raised by the introduction of the new machinery; only 5 percent said that their work pace was reduced. Those who had a job change or transfer only reported much less frequently that increased speed was required of them; and such experiences were almost balanced by cases of a more leisurely work pace.

While having to work faster is disliked, having to exert less physical effort is appreciated and technological advance often does lighten the burden of physical work. Nearly one-half of those experiencing a machinery change reported that they had to work less hard physically after the change, and only 15 percent reported the opposite. For job changers there is on balance no clear-cut change in the physical work burden. The answer "more physical work" is about as frequent as the answer "less physical work." However, it may be inferred from the low correlation and beta coefficients in Tables 6-4 and 6-5 that many workers were not close enough to the tolerable limit of speed and physical exertion before the machine change for changes in these job characteristics to have any major bearing on job satisfaction.

More important (and perhaps surprising to some) is the finding that 60 percent of those who experienced a change in machine technology felt that the work was more interesting after the change than before, while only 5 percent felt that it was less interesting. Indeed, machinery changes increase job interest to about the same extent as do job changes, satisfying a high standard of comparison. The judgment—"The work is more interesting"—represents the single most frequently reported consequence of technological change. We noted earlier that interest is a major determinant of job satisfaction, much more important than the burden of speed and physical work. Perhaps change and variety, of themselves, enhance job interest. However, not too much should be made of this explanation, since many of the job and equipment changes under study here occurred 2 to 5 years ago, enough time for the novelty to have faded out. Young and middle-aged people, the well-educated and the well-paid were more likely than others to find the work more interesting after the equipment or job change, but the differences are minor. In all, there seems to be very little validity in the stereotype which sees mechanization and automation as making jobs increasingly dull, requiring nothing but "robots" to interact with competent and self-sufficient equipment. Although such cases of automation do occur, they seem to be rare in the overall picture of work change.

One reason why work seems to become more interesting is that technological change makes increasing demands on people's faculties (other than mere physical stamina). Among those who experienced a change in machine technology it was reported by 53 percent that as a result their job requires more skill, by 49 percent that the job provides greater opportunity to learn new things, by 48 percent that it enables them to plan, use judgment and initiative to a greater extent, and by 34 percent that it gives them more scope to organize and influence their work. Only 7 percent, or fewer, of people with machinery change answered "less" to any of these questions. All four of these job characteristics are correlated not only with job satisfaction but also with job interest (the simple correlations with job interest are .33, .44, .42 and .36 respectively).[9] By and large technological change enhances the challenge of the job (as measured by these four dimensions) nearly as much as do job changes and transfers.

Two other job characteristics measured in this study reflect the challenge of the job—whether errors are more or less serious and whether the worker has more or less influence on the quality of the product.[10] Both characteristics are strongly correlated with the perceived need for planning and judgment and the perceived opportunity to organize and influence one's own work, although they are less strongly related to job satisfaction than either of these variables. Table 6-6 shows that 31 percent of workers with only machine changes feel that errors are potentially more serious; 29 percent feel that they have more influence on product quality. The opposite answer was in both cases much less frequent. Most people replied "no change."

Finally, we turn to three job characteristics which pertain to the work environment. People who experienced a change in machine technology most frequently saw "no change" in these factors. When questioned about such factors as light, heat, noise, work space, and ventilation, 20 percent reported a change for the better in their work surroundings and 10 percent a change for the worse. Not surprisingly job changers perceived more changes in both directions, and on balance favorable changes considerably outweighed unfavorable ones. Danger of physical injury is negatively correlated with improvements in the work environment, as might be expected. Danger of physical injury is seen as being decreased by mechanization and automation in 25 percent of cases and increased in 15 percent. The figures for job changers are of the same order of magnitude. Twenty percent of people with a machine change felt that the change had increased their chance to talk, while 12 percent felt that it had reduced it. We saw earlier that on most jobs, whether involving the use of machines or not, people feel quite free to talk. Therefore there is relatively little room for improvement. The fact that most work environments are fairly pleasant and free of danger of physical injury should

---

[9]See the correlation matrix (Table 6-11).
[10]This last characteristic was measured for production workers only.

also explain the relatively low frequency of favorable changes of this kind and the weak subjective reaction to these factors.

In all, it appears that technological change typically alters or expands the content of people's jobs in ways which make them feel that the job is more demanding. This is not only true for researchers who through computers gain access to much greater amounts of information, or technicians who man the automated control system of a factory. For production workers more automatic machines often mean that less time and attention is required to operate a particular piece of equipment. The worker's job can then be enlarged by asking him to operate more machines, to handle repairs, or to perform other duties formerly handled by someone else. High speed, integrated equipment raises the flow of output and demands more alertness of the individual worker, new ways of organizing his work flow, and often more coordination between workers.[11] Some managements have deliberately tried to increase job interest and job demands by rotating the work force among several automated tasks. As Mann concludes on the basis of three careful empirical studies:

"It is highly significant that many of the workers in these new plants are finding their jobs not only more demanding, but providing greater variety and interest, more challenge, and increased opportunity to learn. With the goals of greater system integration and efficiency, the design engineer may have accomplished what the industrial engineer failed to do - provide the worker with an environment of tasks appropriate to the skills of a human."[12]

A few examples from the survey of ways in which new or improved equipment made the job more challenging and more satisfying, will help to illustrate the diversity of this type of occurrence:

A 41-year old shipping clerk works for the state liquor commission. He checks whiskey into and out of storage and loads and unloads it. This work was done by hand until 2½ years ago, when the commission got electric fork lifts. Their efficiency meant much less physical work, and 50 percent fewer workers were needed. A larger volume of business as well as the equipment change led to a more demanding job for this shipping clerk. Formerly in charge of only storage, he now manages

---

[11]Evidence of job enlargement also appears in a number of case studies. See for example, Herman J. Rothenberg, "Adjustment to Automation in a Large Bakery," U. S. Department of Labor, Bureau of Labor Statistics, Bulletin No. 1287, *Impact of Automation,* 1960, pp. 84-87; Edward B. Jakubauskas, "Adjustment to an Automatic Airline Reservation System," *Ibid.,* pp. 93-95. Both studies also found cases of downgrading, but upgrading seems to be the more frequent pattern.

[12]Floyd Mann, "Psychological and Organizational Impacts," in John Dunlop ed., *Automation and Technological Change,* The American Assembly, Prentice-Hall, 1962, pp. 51-52. See also Walker, *op. cit.,* pp. 190-217.

both the shipping and storing of his own set of orders. More initiative and planning are needed: "They don't supervise as closely as long as you utilize space well and do a good job. You have to figure out just what you are able to get into a certain area and this requires planning."

A 49-year old physical therapist got new diathermy and automatic traction machines which save her a lot of physical effort. There is less need for planning and mistakes are less serious, because "safety features and timers in the automatic device mean there is less need to be as observant of each and every patient. The machines are pretty well automatic so that there is less possibility of injury to the patient." At the same time, her job has become broader and more challenging. "I can do more variety of work and can handle more patients at one time because I don't have to be constantly attentive."

A 48-year old man who repairs radios and stereo equipment noted that the introduction of transistors, miniaturized components, and stereo by manufacturers meant that he had to learn how each new product worked, and had to change his testing and repair equipment. This required more skill, and made for more challenging and interesting work: "I have to learn constantly and keep up with new trends. Every new piece of electronic equipment means a change in my job."

A 59-year old plant manager of a glass factory reported that the change-over from hand cutting to mechanical glass cutters led to an increase in production. His responsibilities increased greatly, especially in the area of coordination. As he put it, "under the old method each individual who cut the glass was responsible, but now there has to be more planning and scheduling so that everything ties together at the right time, such as the right size boxes ready for the size glass that is to be cut." He considers his job more interesting since the change-over, and especially enjoys the challenge of solving new problems that occur.

Clearly, the scope of each of these jobs was altered by the introduction of some new piece of equipment; greater machine efficiency or productivity enabled each worker to take on a greater variety of tasks. As a result of these new responsibilities, the job became more challenging and interesting.

D. Group Differences In Attitudes Toward
Job Challenge

It has become evident that in the American labor force as a whole the demands or challenges of a job are major determinants of job interest as well as of job satisfaction. In order to assess the role of this complex of factors among different subgroups of the population, it was convenient to construct a scale which would measure change in job challenge using six criteria simultaneously. Each respondent who had an equipment or job change was given a score of +1 for the answer "more" to each of six pertinent questions, and a score of -1 for the answer "less." The questions used as components of the Job Demands Scale were:

1. More or less skill required?
2. More or less speed required?
3. More or less opportunity to learn new things?
4. More or less need for planning, judgment, or initiative?
5. More or less influence in organizing the work?
6. Errors more or less serious?

The points were summed for each respondent, who could receive a total score ranging from -6 to +6. The scale makes possible the ranking of respondents according to the degree to which their jobs became more or less demanding after the job or equipment change; people with negative scores had more decreases than increases in the demands of their jobs, those with low positive scores had on balance a small increase, while respondents with high positive scores presumably experienced the greatest increase in challenge.

The Job Demands Scale is closely related to changes in job satisfaction and job interest as indicated in Table 6-7. In the labor force as a whole a marked association appears between greater job challenge and increased job satisfaction. Less than half of those for whom job demands declined and 55 percent of those seeing no change were more satisfied with their jobs, while almost 80 percent of those at the upper end of the Job Demands Scale reported more satisfaction.

Greater job challenge is even more strikingly related to heightened job interest. Nearly every person with five or six increases in job demands reported that his job had become more interesting. Only 31 percent of those with a decrease in job challenges did so; in fact 41 percent of the latter group said their job was more monotonous.

One would expect that many young people as well as highly educated, professional and technical workers, managers, and businessmen would prefer challenging jobs. The question is whether the observed relation between job challenge and job satisfaction extends to other groups in the labor force. What about less educated or older people or those in blue-collar occupations? The question is partially answered by multivariate analysis. Table 6-8 indicates that among machine changers, job challenge still enhances job satisfaction after interrelationships with socio-economic factors (here the basic set of predictor variables) have been allowed for. In fact, the adjusted coefficients do not show a greatly different pattern from the unadjusted coefficients.

TABLE 6-7

RELATIONSHIP OF JOB DEMANDS SCALE TO CHANGE IN SATISFACTION AND INTEREST
(Unadjusted data)

Change in:	Job demands[a]				
	Decline	No change	1-2 increases	3-4 increases	5-6 increases
Satisfaction					
More	44%	55%	61%	75%	77%
Same	24	31	24	16	13
Less	26	8	10	5	4
Not ascertained	6	6	5	4	6
Total	100%	100%	100%	100%	100%
Interest					
More	31%	40%	64%	84%	92%
Same	27	52	27	11	4
Less	41	8	8	5	4
Not ascertained	1	*	1	*	*
Total	100%	100%	100%	100%	100%
Number of cases	(160)	(255)	(290)	(319)	(205)

* Less than 0.5 percent.

[a] Includes only those workers who have experienced a machine change or job change or both.

TABLE 6-8

RELATIONSHIP OF JOB DEMANDS TO JOB SATISFACTION**

Job demands scale[c]	More satisfied[a]			Less satisfied[b]		
Mean:	63.5%			9.2%		
	Number of cases	Deviations from mean		Number of cases	Deviations from mean	
		Unadjusted	Adjusted		Unadjusted	Adjusted
Decline	79	-18.0%	-14.0%	160	+17.1%	+12.8%
No change	94	-5.0	-3.1	255	-1.7	-1.4
1-2 increases	141	-2.6	-2.2	290	+0.8	+0.6
3-4 increases	165	+10.4	+7.1	319	-4.5	-3.5
5-6 increases	108	+5.0	+5.1	205	-5.3	-3.7
Beta coefficient	.14			.18		
	$R^2 = 10.4\%$			$R^2 = 13.5\%$		

** Also included in each regression were: income change over the past 5 years, kind of work change in past 5 years, occupation, age, education, and 1966 income from worker's job.

[a] Includes only those with a machine change.

[b] Includes those with a machine and/or job change.

[c] The -6 to +6 Job Demands Scale described on page 27 was condensed to these five categories for multivariate analysis.

TABLE 6-9

RELATION BETWEEN CHANGE IN JOB SATISFACTION AND IN VARIOUS JOB
CHARACTERISTICS RANKED BY SIZE OF THE SIMPLE CORRELATION COEFFICIENT
FOR EDUCATION AND AGE SUBGROUPS[a]

Change in:	Education		Age	
	Less than high school	High school degree or more	Under age 45	Age 45 or older
1. Income	1	1	1	1
2. Opportunity to learn	2	2	2	5
3. Interest	3	3	3	2
4. Chance for advancement	4	4	4	4
5. Planning, judgment, initiative required	5	5	5	6
6. Skill required	6	6	6	3
7. Own influence on work	7	7	7	7
8. Steadiness	9	9	9	9
9. Seriousness of errors	14	8	8	10
10. Chance to talk	10	10	12	8
11. Pleasantness of physical surroundings	13	11	11	12
12. Influence on quality of production	8	12	10	11
13. Closeness of supervision	12	14	13	16
14. Amount of paper work	11	16	14	14
15. Danger of personal injury	15	15	15	15
16. Speed required	17	13	16	13
17. Physical effort required	16	17	17	17
Number of cases	(385)	(825)	(876)	(347)

[a]Includes only those workers who have experienced a machine and/or job change.

Table 6-9 explores the problem further by focusing specifically on two groups for which one might doubt the positive relationship between job challenge and job satisfaction—workers in the older age brackets and those who have not completed high school. The simple correlations between changes in various job characteristics and changes in job satisfaction (Table 6-4) were obtained separately for two-age and for two-education groups.[13]   The four

---

[13]A more detailed breakdown, separating out people over age 55 or those with less than 8 years of schooling, would have resulted in an undesirably small number of cases.

columns of Table 6-9 present the correlation coefficients for each subgroup ranked by size, 1 signifying the closest relation to job satisfaction and 17 the least. The list of job characteristics is the same as in Table 6-4, in descending degree of correlation with job satisfaction for the whole sample. Table 6-9 exhibits a remarkably similar rank ordering for the four groups.[14] That is, job characteristics have much the same relative importance for job satisfaction among younger and older workers as well as among more and less educated workers. The variables which measure perceived job challenge rank as high as determinants of job satisfaction among people who have not completed high school and among those over age 45 as among other members of the labor force. One notable exception is the chance to learn. For people over age 45, greater skill requirements contribute more, and chance to learn contributes less to job satisfaction than among other groups—hardly an important difference! A chance to talk also seems to be somewhat more important for older than for younger workers. Interestingly, chance for advancement and steadiness rank the same (4th and 9th respectively) in all four groups. In all, the table reveals surprisingly few differences among groups in the influence of various job characteristics on job satisfaction.

A third and final approach to the question consisted of looking at two "deviant" groups: (1) those for whom lower or unchanged job demands were associated with *increased* job satisfaction instead of reduced job satisfaction, and (2) those for whom greater job demands were associated with unchanged or reduced job satisfaction. Do these "deviant" groups have any distinctive socio-economic characteristics? The two groups are compared in Table 6-10 with the two groups representing the "dominant" pattern: (1) workers reporting greater job demands and more satisfaction and (2) workers with lower or unchanged job demands and reduced job satisfaction.

Some interesting differences between the four groups are revealed by Table 6-10. As expected, the group reporting both increased challenge and increased satisfaction (column 1) is younger and more highly educated than the others; 53 percent are under age 35, and 39 percent have at least some college training. Professional, technical, and managerial people are somewhat more frequent in this group than among all others. The second group which conforms to the "dominant" pattern, those reporting the same or lower job demands and less job satisfaction (column 2), contains a higher proportion of older and less educated people. About one-half of them are operatives, service workers, and laborers. These findings again indicate that job challenge is a prerequisite for job satisfaction even in the less educated and blue-collar groups. New equipment or jobs which fail to make greater demands on the capabilities of these workers fail to arouse feelings of increased job satisfac-

---

[14]The correlation coefficients are higher for each of the 4 subgroups than for the whole sample, in part because the subgroups are more homogeneous.

TABLE 6-10

MAJOR SOCIO-ECONOMIC CHARACTERISTICS OF JOB DEMANDS - SATISFACTION GROUPS[a]

(Unadjusted data)

	Greater job demands, more satisfaction	Same or lower job demands, same or less satisfaction	Greater job demands, same or less satisfaction	Same or lower job demands, more satisfaction
Age				
Under age 25	19%	10%	13%	15%
25-34	34	24	20	36
35-44	26	21	28	18
45-54	14	25	21	20
Age 55 or older	7	20	18	11
Total	100%	100%	100%	100%
Education				
0-7 grades	5%	11%	4%	9%
8-11 grades	20	37	30	30
High school degree	36	32	33	32
Some college	21	11	17	20
B.A. degree or higher degree	18	9	16	9
Total	100%	100%	100%	100%
Occupation				
Professional and technical workers	19%	9%	16%	11%
Managers, officials	10	2	6	6
Proprietors, self-employed businessmen	3	1	2	1
Clerical and sales workers	22	17	25	18
Craftsmen, foremen	19	18	19	17
Operatives	16	31	18	28
Service workers	5	13	6	9
Laborers - nonfarm	4	4	3	6
Farmers and farm workers	2	5	5	4
Total	100%	100%	100%	100%
Income change				
Increase	81%	58%	73%	73%
No change	8	24	11	13
Decrease	3	16	12	9
Not ascertained	8	2	4	5
Total	100%	100%	100%	100%
Number of cases	(573)	(177)	(199)	(210)

[a]Includes only those workers who have experienced a machine and/or job change.

tion. Many are obviously not content with routine unchallenging work, even if it happens to be steady and provide fairly pleasant surroundings.

The first "deviant" group (column 3) which combines greater job demands with *less* satisfaction is somewhat older and less educated than the corresponding dominant group, which combines greater job demands with *more* satisfaction (column 1); the main point however is that the socio-economic differences are not large. Besides, the deviant group had fewer income increases, a situation which tends to depress job satisfaction. The second "deviant" group (column 4), which is more satisfied despite lower or unchanged job demands is *not* of lower socio-economic status than its "dominant" counterpart (column 2). On the contrary it is somewhat younger, better educated, and more often in high-status occupations—characteristics which are conducive to optimistic expectations. Again, these socio-economic differences are quite small. A more sizable difference appears in the income change pattern of the two groups. The greater job satisfaction of the "deviant group" despite the lack of new challenge of the work, is associated with a higher frequency of income increases. This group seems to consist to some extent of people who are willing (perhaps temporarily) to disregard the failure of the job to provide new challenge after the equipment change, since the new work provides them with a higher income. No doubt there are personality differences and differences in expectations as well.

*   *   *

The analysis in this chapter may be summarized by stressing three important findings. First, besides income change, the major determinants of both increased and decreased job satisfaction among people who have experienced a change in equipment are those job characteristics which relate to job challenge; more demanding jobs mean increased job satisfaction. Second, equipment change seems to raise the challenge of a job much more often than it reduces it. It tends to enlarge the job in a non-physical sense, while the physical work burden is often reduced by technological advance. Third, the tendency of increased job challenge to generate increased job satisfaction extends to all major socio-economic groups. It is *not* confined to the young, well-educated, or to white-collar workers. These findings do not answer the question whether automation in a narrowly defined sense increases work challenge, since many of the equipment changes studied here probably did not involve the substitution of logical for other controls, but were simply an increase in mechanization (i.e. a movement along the Automation Scale, but not necessarily into one of the top categories). There is evidence that workers who after an equipment change worked with logically controlled equipment were about as likely to feel increased job satisfaction as others who have experienced an equipment change.

The bleak prospects envisaged by Adam Smith nearly 200 years ago regarding the impact of machines on the work force have not materialized to date. Yet, similar fears are expressed today regarding future developments. Our findings suggest that given the present pace and structure of technological change these misgivings about the future would seem to be unwarranted.

TABLE 6-11

CORRELATION MATRIX OF CHANGE IN SATISFACTION AND CHANGE IN JOB CHARACTERISTICS[a]

Change in:	Job satis-faction	1	2	3	4	5	6	7	8	9	10	11	12	13	14	15	16
1. Income	.30																
2. Opportunity to learn	.27	.08															
3. Interest	.27	.10	.44														
4. Chance for advancement	.25	.10	.39	.29													
5. Planning, judgment, initiative required	.20	.08	.47	.42	.27												
6. Skill required	.19	.14	.43	.33	.21	.40											
7. Own influence on work	.15	.03	.34	.36	.20	.47	.24										
8. Steadiness	.12	.03	.24	.20	.30	.15	.11	.15									
9. Seriousness of errors	.11	.09	.30	.21	.18	.31	.30	.19	.10								
10. Chance to talk	.10	.03	.17	.22	.16	.16	.10	.14	.13	.06							
11. Pleasantness of physical surroundings	.08	.03	.14	.20	.10	.17	.10	.17	.08	.06	.17						
12. Influence on quality of production	.06	-.01	.27	.11	.16	.19	.18	.14	.06	.07	-.05	.04					
13. Closeness of supervision	.03	-.02	.07	.02	.20	-.01	.12	-.14	.06	.11	-.03	.02	.16				
14. Amount of paper work	.03	.03	.16	.13	.10	.19	.18	.15	.08	.16	.04	.07	.02	.05			
15. Danger of personal injury	.02	-.01	.12	-.01	.04	.07	.11	-.02	.03	**.12**	-.02	-.18	.07	.01	.03		
16. Speed required	-.01	.05	.03	-.01	.05	.07	.17	-.03	.02	.13	-.03	-.07	.08	.15	.05	.34	
17. Physical effort required	-.05	.02	-.03	-.03	**-.02**	.03	.05	-.02	.01	.03	.00	-.07	.00	-.06	.00	-.01	.22

[a]Includes only those workers who have experienced a machine and/or job change.

/

Chapter 7

FURTHER ANALYSIS OF CASES WHO MADE A POOR ADJUSTMENT TO TECHNOLOGICAL CHANGE

The preceding chapters have shown that a large majority of workers adjust easily to technological change. For many, technological change means an advance in their careers and a new challenge. There is no evidence that working with machines or working with machinery which has been improved technologically enhances the chance that a worker will be unemployed. Attitudes toward machines and even toward automation are predominantly favorable among American workers. Nevertheless, for a small minority of workers these favorable generalizations do not hold.

People who experienced no change in machine technology were no doubt in a more static work situation than those who experienced technological change. That is, this "no change" group reported constant incomes and expressed neutral attitudes somewhat more frequently than others. The group which was affected by technological change was more likely to experience changes in income—either up or down—and to have affective reactions toward the machines they work with—either positive or negative. Hence, even though favorable experiences and attitudes were relatively frequent among workers who underwent a change in machine technology, unfavorable experiences and attitudes also occurred with above-average frequency.

The minority for whom technological change has meant a setback, in terms of pay, career, steadiness of employment, or job satisfaction, is of particular interest for public policy. These are the people who are the potential beneficiaries of programs to alleviate the impact of technological change. A thorough look at these cases should provide a useful background for policy decisions.

The special analysis of the small group for whom technological change created difficulties will proceed in two steps. To start with, on the basis of eleven criteria a simple scale is constructed which ranks workers from those who seem to have made the poorest adjustment to technological change to those who have made the most successful adjustment. For convenience of expression we shall call this the "Adjustment Scale." The construction of two versions of this scale is described in Section A. In Section B, the distribution of workers along the Adjustment Scale is analyzed by socio-economic characteristics and by the nature of the change which they experienced. Some of the eleven components of this scale were related individually to the same variables in earlier chapters. The purpose of Section B is to bring into sharper focus those workers who gave evidence in reply to *several questions* of having reacted poorly to technological change. A related purpose is to compare this group of workers with those who had particularly favorable experiences and attitudes. Section C supplements the statistical approach by a series of case studies. It summarizes the interview with a dozen workers who made a particularly poor adjustment (in terms of the Adjustment Scale) to a change in machine technology.

A. The Adjustment Scale

The Adjustment Scale was constructed for all people who experienced a change in machine technology, with or without an accompanying job change. People who changed jobs and used the same equipment on the new job are excluded from the analysis in this chapter, since the scale is designed to measure adjustment to new machines rather than a new job.

Two versions of the Adjustment Scale were constructed. The first version, called here Adjustment Scale A, registers only the frequency of negative responses; that is, it scores the number of indications that the worker made a poor adjustment to technological change. The following eleven criteria were used to compute a score for each worker.

    (1) Earning the same or a lower income than 5 years ago

    (2) Two or more spells of unemployment in the past 5 years

    (3) At least one spell of unemployment lasting a month or more

    (4) Being less satisfied with one's job than 5 years ago

    (5) Finding the work less interesting than before the machine (and job) change

    (6) Having less steady work than before the machine (and job) change

   (7) Seeing less chance for advancement than before the machine (and job) change

   (8) Saying that automation is a bad thing for people in one's line of work

   (9) Saying that one's equipment is like "a foe"

  (10) Saying that the job is "drudgery"

  (11) Describing only unfavorable aspects of one's job

Each of these responses was given a score of 1. The most unfavorable possible score would then be 11, the most favorable score 0. In fact, as Table 7-1 indicates, only 2 percent of workers who experienced a change in machine technology had a score of 6 or above; and less than 4 percent had a score above 4 (more than four unfavorable replies out of a possible eleven). Nearly half of all workers who underwent a change in machine technology scored 0 on Adjustment Scale A (not a single unfavorable response).

The second version, called Adjustment Scale B, registers positive *and* negative responses and therefore represents a *net* score. All favorable responses are scored +1, all negative responses -1, and neutral responses are scored 0. For example, a worker who experienced an income decline, but felt that his chances for advancement had improved would be scored 1 on Adjustment Scale A (assuming that all other responses were neutral); he would be scored 0 on Adjustment Scale B, where the positive response would offset the negative one. The negative criteria are the same as for the A scale. The criteria used to identify successful adjustment are for the most part the opposite of the negative criteria listed above:

   (1) Earning a higher income than 5 years ago

   (2) Having been promoted during the past 5 years

   (3) Being more satisfied with one's job than 5 years ago

   (4) Finding the work more interesting than before the machine (and job) change

   (5) Having more steady work than before the machine (and job) change

   (6) Seeing more chance for advancement than before the machine (and job) change

   (7) Saying that automation is a good thing for people in one's line of work

   (8) Saying that one's equipment is like "a friend"

   (9) Saying that one "enjoys" one's job

  (10) Describing only favorable aspects of one's job

Potentially the B scale ranges from +10 to -11. Table 7-1 shows that only 7 percent of workers who experienced a change in machine technology had a

TABLE 7-1

DISTRIBUTION OF WORKERS ACCORDING TO ADJUSTMENT
SCALES A AND B

	Adjustment Scale A	Distribution of workers[a]
Best adjustment	0	47%
	1	24
	2	14
	3	8
	4	4
	5	1
	6	1
Worst adjustment	7 or more	1
	Total	100%

	Adjustment Scale B	Distribution of workers[a]
Best adjustment	8, 9, or 10	12%
	6 or 7	25
	4 or 5	31
	2 or 3	17
	0 or 1	8
Worst adjustment	negative score	7
	Total	100%
	Number of cases	(587)

[a]Includes only those workers who have experienced an equipment change.

negative score on balance, while 37 percent had a score of +6 or more. Thus Adjustment Scale B is dominated by the frequency of favorable responses, reflecting the fact that on the whole positive reactions outweigh negative ones.

B. Statistical Analysis of Workers Who Made a Poor Adjustment

How well a worker adjusts to changes in the equipment he works with depends on the circumstances surrounding the equipment change and also on his socio-economic characteristics. These circumstances and characteristics tend to be interrelated. Hence they should be analyzed by multivariate statistical techniques. As in Chapters 5 and 6, the particular technique employed was the Multiple Classification Analysis (MCA).[1] The dependent variables were in turn Adjustment Scales A and B to allow for the possibility that the factors which make for a poor adjustment may not be simply the reverse of those which make for a good adjustment. It turned out, however, that the two sets of results closely paralleled each other. Hence in the tables which follow only the statistical analysis for Adjustment Scale A is presented, which measures indications of a poor adjustment and is the major focus of the analysis. Occasional differences between the two sets of results are noted in the text. The grand mean for the number of indications of a poor adjustment is 1.13, with the observed values ranging from 0 to 9. Positive deviations from the mean imply additional indications of poor adjustment, negative deviations a better than average adjustment. First a basic set of five independent or predictor variables was related to the Adjustment measure—occupation, age, income level, education, and whether or not the equipment change was connected with a job change or transfer. Additional explanatory variables were added one at a time in order to test their relevance to adjustment, after allowing for other interrelated characteristics. The analysis was limited to people who experienced an equipment change during the past 5 years. Table 7-2 presents the adjusted and unadjusted deviations for the five basic predictor variables as well as the respective beta coefficients,

Judging by the beta coefficients, occupation is the most important determinant of the kind of adjustment which people make to technological change, with operatives and service workers experiencing appreciably more difficulty than all other groups. Professional and technical personnel together with the managerial and administrative group make the best adjustment. Clerical and sales workers are in an intermediate position, as are craftsmen and foremen. That people react somewhat more favorably to technological change in the office than in the factory is also suggested by the industry breakdown shown in Table 7-3 (unadjusted data). Those who are employed in manufacturing and construction give relatively frequent indications of a poor reaction to technological change. On the other hand, people employed in finance, real estate, and business services score favorably on the Adjustment Scale. This is an industry group in which office automation was particularly extensive.

That poor adjustment to technological change occurs relatively frequently at lower levels of income and education is hardly surprising and requires no comment (Table 7-2). The further finding that older people make a poorer adjustment than young people is also expected, yet one might have looked for a stronger and more consistent relationship. The interviews indicate that when older people experience changes in machine technology on a job

---

[1]For a description of the program see Appendix IV, and also the briefer explanation in Chapter 5, Section C, where the program was first used.

TABLE 7-2

RELATIONSHIP OF BASIC DEMOGRAPHIC AND ECONOMIC VARIABLES
TO ADJUSTMENT SCALE A[a]

Mean Number of Indications of Poor Adjustment: 1.13

	Number of cases	Deviations from mean		Beta coefficient
Occupation		Unadjusted	Adjusted	
White collar	131	-.63	-.40	
Clerical and sales workers	119	-.28	-.27	
Craftsmen, foremen	115	-.35	-.32	
Blue collar	216	+.72	+.57	
				.29
Kind of change				
Machine change only	224	-.53	-.44	
Change in employer and machine change	292	+.40	+.31	
Transfer and machine change	71	+.02	+.10	
				.23
Age				
Under age 25	107	-.04	-.33	
25-34	179	+.20	+.11	
35-44	144	-.25	-.11	
45-54	95	-.09	+.14	
Age 55 or older	61	+.19	+.30	
				.13
1966 income from worker's job				
Less than $4,000	175	+.38	+.11	
$4,000-5,999	131	+.20	+.09	
$6,000-9,999	193	-.20	-.06	
$10,000 or more	86	-.66	-.25	
				.10
Formal education				
0-7 grades	35	+.70	+.26	
8-11 grades	154	+.45	+.14	
High school degree	201	-.23	-.17	
College[b]	190	-.28	-.01	
				.10

$$R^2 = 20.2\%$$

---

[a] Includes only those workers who have experienced an equipment change; number of cases = 587.

[b] This group includes all those with college experience of at least one year.

TABLE 7-3

ADJUSTMENT SCALE A RELATED TO INDUSTRY[a]

(Unadjusted data)

Adjustment Scale A		Industry						
		Construction	Manufacturing	Transportation, communication, utilities	Trade- wholesale and retail	Finance, business services	Repair, personal health, education and other services	Govern- ment
Best adjustment	0	37%	39%	52%	44%	70%	58%	53%
	1	23	20	22	31	20	25	33
	2	14	18	12	15	2	10	5
	3	14	11	10	7	3	2	5
Worst adjustment 4 to 9		12	12	4	3	5	5	4
Total		100%	100%	100%	100%	100%	100%	100%
Number of cases		(43)	(217)	(50)	(68)	(40)	(81)	(43)

[a]Includes only those workers who have experienced an equipment change.

which they have held for some years, seniority rights sometimes cushion them against layoffs or downgrading. The relationship between age and Adjustment Scale B is more pronounced. The reason is that favorable attitudes and experiences (which are included in the B scale, but not in the A scale) are more commonly reported by younger than by older people.[2]

There is a slight tendency for white workers to encounter fewer difficulties in adapting to technological change than nonwhite workers; but the difference nearly disappears in the adjusted coefficients. Thus what appears to be a racial difference is no more than a reflection of the occupational, educational, and income characteristics of Negroes, which of themselves would imply a relatively poor adjustment to technological change. Ability to adjust to technological change also shows no appreciable difference between men and women.

Next we turn from personal characteristics which may affect adjustment to characteristics of the new job and circumstances surrounding the change—over. Table 7-2 indicates that people who had to change equipment *and* jobs (or sections within the same company) more often had difficulties than those who could stay on the same job when new equipment was introduced. Indeed after occupation, this variable shows the strongest relationship to Adjustment Scale A. However, the difference diminishes when positive criteria are also taken into account (Adjustment Scale B). Those who changed jobs tended to experience some favorable changes along with unfavorable ones, while neutral (or "no change") responses were more frequent among those who remained on the same job.

A further breakdown of the job changers into those whose former employer was cutting back production and those where this was not the case shows that the former group had more negative attitudes and experiences than the latter, i.e. ranked lower on the Adjustment Scale (Table 7-4). For whatever reason the employer was cutting back, the pressure to leave the former job was undoubtedly more urgent for people who were employed in declining firms. People who said they changed jobs *because of* equipment changes also scored relatively low on the Adjustment Scale. People who are under some pressure to find a new job may have less opportunity than others to locate employment which is really congenial to them and which is steady and well—paying. This is one explanation for their poorer adjustment. A reading of the interviews suggests an additional explanation. There is some tendency for people with marginal work qualifications, i.e. people who have a weak bargaining position in the labor force, to have to accept jobs which are unsteady, seasonal, or in declining firms. In a sense they are forced to move from one

---

[2]Young people generally give more optimistic replies to survey questions which measure attitudes.

TABLE 7-4

ADJUSTMENT SCALE A RELATED TO CIRCUMSTANCES
UNDER WHICH WORKERS LEFT FORMER EMPLOYER[a]

(Unadjusted data)

Adjustment Scale A	Former employer was cutting back production	Former employer was not cutting back production
Best adjustment 0	24%	39%
1	29	26
2	16	17
3	13	10
Worst adjustment 4 to 9	18	8
Total	100%	100%
Number of cases	(62)	(303)

[a]Includes only those workers who have experienced an equipment change.

TABLE 7-5

NUMBER OF EMPLOYERS IN PAST 5 YEARS BY ADJUSTMENT SCALE A[a]

(Unadjusted data)

Adjustment Scale A	Number of employers in past 5 years			
	One	Two	Three or four	Five or more
Best adjustment 0	64%	37%	35%	17%
1	18	26	26	30
2	8	19	16	23
3	5	9	10	15
Worst adjustment 4 to 9	5	9	13	15
Total	100%	100%	100%	100%
Number of cases	(217)	(134)	(102)	(40)

[a]Includes only those workers who have experienced an equipment change.

job to another. Table 7-5 should probably be interpreted in these terms. It suggests that the more employers a worker has had in the past 5 years, the poorer is his adjustment to technological change.

Did those workers who—*after* the machine change—worked with highly automated equipment make a better (or worse) adjustment than others? According to Table 7-6, there is some (not very strong) tendency for members of the labor force who do not operate equipment[3] and those who operate the simpler kinds of manually guided equipment to give more indications of a poor adjustment to technological change than those who work with more mechanized or automated equipment. The unadjusted deviations suggest that people who work with logically controlled equipment (the highest category on the atuomation scale) make a much more successful adjustment than others. However, this difference disappears when their superior educational and occupational status is taken into account. Equipment with which people work indirectly has less influence on the kind of adjustment they make to technological change than equipment they operate themselves, with one exception: Workers who have indirect contact with equipment under fixed mechanical control (most of those who work on an assembly line or conveyor fall into this category) *do* make a somewhat poorer adjustment than others.

Finally we may look at the incidence of poor adjustment in relation to the demands created by the new equipment (and sometimes job). In Table 7-7, Adjustment Scale A and B are related to the Job Demands Scale described in Chapter 6. It appeared in Chapter 6 that jobs which became increasingly demanding in terms of skills, opportunity to learn, need for planning and judgment, and chance to organize one's own work also tended to become more satisfying. The upper part of Table 7-7 shows that people who felt that after the equipment change their work was less demanding, than it was before the change, were considerably more likely than others to have adjusted poorly to the new job situation. This is true even after allowing for socio-economic characteristics of workers and the circumstances surrounding the equipment change. Indeed the relation between Job Demands and Adjustment Scale A is somewhat stronger (as measured by the beta coefficients) than the relation between either income, age, or education and the same Adjustment Scale. The relation between adjustment and job demands appears even closer in the lower part of Table 7-7. Adjustment Scale B is utilized there, which like the Job Demands Scale has a positive as well as a negative dimension. Evidence of a good adjustment rises with greater job demands. The beta coefficient is .40. The direction of causation is not entirely unambiguous. Yet it is clear that many American workers do see a more demanding job as a better job.

---

[3]These people either had a change in equipment with which they have only indirect contact; or as a consequence of the machine (and job) change, they no longer operate equipment.

TABLE 7-6

RELATIONSHIP BETWEEN AUTOMATION LEVEL OF EQUIPMENT
AND ADJUSTMENT SCALE A[a]

Mean Number of Indications of Poor Adjustment: <u>1.13</u>

Automation level of equipment operated directly	Number of cases	Deviations from mean		Beta coefficient
		Unadjusted	Adjusted	
Numerical, tape, computer or other logical control	41	-.69	-.08	
Fixed mechanical control	273	+.02	+.03	
Powered multi-system, manual main control	113	-.01	-.25	
Manual control, operator powered, or powered single-system	76	+.47	+.31	
No equipment	77	-.08	+.10	
				.15
Automation level of equipment used indirectly				
Numerical, tape, computer or other logical control	72	-.59	-.08	
Fixed mechanical control	130	+.07	+.14	
Powered multi-system, manual main control	49	-.21	-.10	
Manual control, operator powered, or powered single-system	b	b	b	
No equipment	324	+.15	-.01	
				.06

$$R^2 = 21.3\%$$

---

[a]This equation includes the five basic predictors in Table 7-2. Only workers with an equipment change are included.

[b]Only 8 cases.

TABLE 7-7

RELATIONSHIP OF JOB DEMANDS SCALE TO ADJUSTMENT SCALES[a]

Mean Number of Indications of Poor Adjustment (Scale A):   1.13

Job Demands Scale[b]	Number of cases	Deviations from mean		Beta coefficient
		Unadjusted	Adjusted	
Decline	79	+1.59	+1.19	
No change	94	-.34	-.20	
1-2 increases	141	-.06	-.04	
3-4 increases	165	-.32	-.27	
5-6 increases	108	-.30	-.23	
				.32

$$R^2 = 28.9\%$$

Mean Net Adjustment Score (Scale B):   3.95

Job Demands Scale[b]				
Decline	79	-1.37	-1.18	
No change	94	-.26	-.19	
1-2 increases	141	-.12	-.07	
3-4 increases	165	+.52	+.41	
5-6 increases	108	+.59	+.50	
				.40

$$R^2 = 28.1\%$$

[a]This equation included the five basic predictors in Table 7-2. Only workers with an equipment change are included.

[b]The -6 to +6 Job Demands Scale described in Section D of Chapter 6 was condensed to these five categories for multivariate analysis.

In sum, a number of personal and situational characteristics seem to influence how a worker fares when his job is affected by changes in machine technology. The following sketch might be used to describe the typical worker who is unfavorably affected by technological change: an operative or service worker who has not completed high school; he is elderly, has a modest income, is forced by his old employer's equipment change or by cutbacks in production to take a job which involves the use of different equipment; after the change he operates no equipment or manually controlled equipment on a production line, and finds the new job unchallenging and undemanding. Before we turn from this abstraction to some of the actual cases, it must be emphasized once more that a large majority of workers in *all* subgroups of the population which were examined adjusted readily to technological change. In *every* group there were only a few who showed signs of a difficult adjustment. Specifically, in none of the subgroups analyzed did the proportion of workers who scored four or more negative criteria on Adjustment Scale A (out of a possible 11) exceed 20 percent.

C. Case Studies of Workers Who Made a Poor Adjustment

Interviews are summarized below for a dozen workers who made a poor adjustment to a change in machine technology. Some stayed with the same employer, some changed jobs *because of* the machine change, some changed jobs for other reasons and had to use different equipment on the new job. In cases where there was a combined job and equipment change, the Adjustment Scale measures the combined reaction to the different equipment and other features of the new job; they cannot be disentangled. All cases described scored at least 5 on Adjustment Scale A, i.e. they met five or more of the negative criteria listed earlier. This was the only principle of selection. For brevity's sake not all interviews in this category are summarized. Those omitted are less interesting because they involve inarticulate respondents or because they relate to unusual circumstances, for instance, some college students working part time, a Cuban refugee, a man just released from the penitentiary.

Case 1:   A construction laborer in Texas, male, white, 60 years old, married with one child under 18.   He grew up on a farm, had 12 years of schooling, including some vocational training in agriculture while in high school; 1966 income $4,000 to $4,999.

For 18 years he was an operator in a crude oil pumping station. It was originally a diesel station. He kept the diesel engines in operation, oiled them, looked after them, and changed the valves on the tanks. Four years ago the company changed to electricity and automated the whole station. It became a "push button job." "One man in Houston can operate everything from Houston to Tampa. So it's cut out lots of people. The whole system is automatic." A hundred or more people were thrown out of work, according to this respondent. His own job was also abolished. He took early retirement at about 1/3 the retirement income he would have received at age 65.   The oil company still pays for his life and health insurance. He was out of work for 4 to 5 months, the only spell of unemployment he has had in the past 5 years. Then he found his present job.

The change-over was company-planned. The workers knew 2 years in advance that it was coming. "We were sitting there watching them set it up. The 5 percent Social Security is what makes them have to do it."

On his present job this man finishes concrete, drives a tractor to dig ditches, helps the carpenters, and "whatever's to be done around a construction job." His major pieces of equipment are a tractor with power parts and an electric saw. Although the company figures that it takes 60 days to break in a man for this type of job, this man received no formal training.

The hours are longer on the new job, and the pay is lower (but he has some retirement income to supplement it). He is less satisfied with his work than 5 years ago. His main complaint is the pay. He says he may change jobs again in order to get more money. Chances for advancement are poor: "Smaller company and I am too old to make a much better hand." The new job is also less interesting: "Well, there's no advancement. It's just something to do." It is more demanding in terms of physical energy and speed, but provides less opportunity to learn, organize one's own work, or plan than the old job did. Still, he calls the new job "enjoyable." He can talk more, the physical surroundings are more pleasant, and he particularly likes the people he works with.

Not surprisingly, his attitude toward automation is unfavorable: "It throws people out of work. . . . Someday I just won't have any work. Machinery will take the place of lots of men. You don't have to pay taxes or Social Security on machines."

Case 2:   A worker on the assembly line of an automobile plant, male, white, 36-years old, married with 4 children, 10 years of schooling in a small town in North Carolina, no further training, no need felt for additional education. He made $7,500-10,000 in 1966.

Up to 3 years ago he drove a delivery truck for an oil company. He left that job for financial reasons and found the automobile company job. There he works on the assembly line, attaching a steel support on the motor by using an air pressure wrench, which is guided manually as it puts on nuts and tightens them. The company trained him for 3 days, but he says that one day would have been sufficient. The work is paced by the production line. "I can talk a little. If I didn't, I would go crazy."

He feels that the new job demands more in the way of physical energy and speed than the former one. It also is more closely supervised than the old. "We have more than enough supervisors and foremen watching what we do. On my old job I made the deliveries by myself." However, his pay was raised when he took the new job. Taking everything into consideration he says that he is about as satisfied on the new job as on the old job. "It gets pretty monotonous. . . . Doing the same thing all day, every day. . . . I do the same thing every day, and I see the same people. I had a different schedule of work on my previous job. But this job pays more, so. . . ." In spite of his dislike for it, he plans to keep the job.

This man describes the machinery he works with as a "foe." In his view "automation causes problems, because so many jobs are taken over by machines."

He says that he has never been unemployed.

Case 3: A 47-year old white man, with 12 years of school. He took a 2-year training program in banking, and used to be a teller many years ago. He earned $7,500 to $10,000 in 1966, and has never been unemployed.

For 15 years he was a sales supervisor for a large drug company in New York City: "I had contact with various customers regarding sales and potential sales increases. I was involved in the introduction of new products also." Then 5 years ago the company acquired a large computer and moved it outside of the city. Employees had no voice in the equipment change. "I had nothing to do with it. Perhaps 'higher-ups' participated, but the general run of employees had nothing to do with it." Many fewer people were needed; some became unemployed, and his own job was abolished. He was transferred (on one year's notice) to a different job.

The new job is in the production planning and inventory control department of the drug company. He uses no equipment directly, but a computer is important for his work. It provides data on past sales for different products in different areas; these statistics are used in reports to estimate future sales, production levels, and inventories.

The company trained him 35 hours on the job in use of computer results, and he read on his own about the operation of the computer.

He dislikes his new job and considers it drudgery; the only good thing about it is "the fact that I do have a job." He thinks it "is dull and boring. The day is longer, and there is no contact with people. Because I am dealing with statistics, it can be most monotonous. I like to deal with people." The crux of his complaint seems to be: "Although I was happy as a salesman, they transferred me to the statistical department, where there is no more selling."

As a result of this unhappy transfer, many things are less satisfactory. According to him, the new job requires less skill, offers less opportunity to learn, and gives him less influence. "I don't work with people, just records." In addition, the physical surroundings are less pleasant, and he can't talk as much, which is frustrating to him.

Although his income was raised, he credits it to seniority rather than the transfer. And future advancement is unlikely because "they don't move up men at my age."

His attitude toward automation is negative, "because by and large it destroys relationships. Basic understanding among personnel and the social aspect of a job are destroyed by it."

He plans no job change, but indicates that additional education "would give me an opportunity to get into another division and away from the computer."

Case 4:   A trimmer with a meat packing company, 48 years old, white, male, married with one child under age 18. He grew up in rural Mississippi, had 8 years of schooling plus a 3-month course in welding, financed under the G.I. Bill. He feels no need for further education or training, made $5,000 to $6,000 in 1966.

This man has had several construction jobs during the past 5 years; one employer moved, another did not have enough work, a third did not pay enough. Finally he went to the meat packing company. He accomplished all these job changes with only one week of unemployment between jobs. The one week was the only spell of unemployment he has had in the past 5 years. His pay was raised when he took the meat packing job, and he is earning more than he was 5 years ago. Also, he is obtaining fringe benefits and seniority rights for the first time.

On his present job he works on a production line. His particular task is to split the back of the cow and pop the kidney. For this he uses an air knife which is guided manually. It took him 2 to 3 days to learn the job.

Aside from his pay, this man is decidedly less satisfied with his work than 5 years ago. The production line was particularly hard to get used to. He has to work faster than on his former job. He used to work outdoors; his new surroundings in the slaughterhouse are decidedly less pleasant. The work is paced by the production line. "If you do your job right, you don't have time to talk." There is little opportunity to learn or to use his own judgment. "What I do now is just automatic. You do the same thing in the same way every day. . . . When you do the same thing every day it is monotonous."

He says he doesn't like his present job at all. He likes nothing about it. "I had rather do another kind of work such as construction." It is possible that he may change jobs again. "I don't know that there would be any advantage with staying. I'd like to be getting on the outside, working out of doors."

Regarding automation: "I don't think it makes any difference actually; I don't think machines could take anybody's place. You always have to have someone to take care of the machines."

Case 5:   A lathe and punch press operator for a tool and die company. Fourteen and a half years with the same employer, white, 32 years old, married with three children, 10 grades of schooling without any vocational or other special training. He made $7,500 to $10,000 in 1966.

This man operates a punch press, lathe, and drill press to make automobile parts. "You control the part and watch it, and the die comes down on the part." There are also conveyors and air feeders which bring the work and take away the scrap and stock.

Two years ago "they moved the machines closer together. The machines now hold four or five instead of two pieces at a time. They put more dies in the press so it would have more parts in it and do the work faster." For him this change means that he has to run more pieces through the machine per hour. "There are too many dies in the press. The equipment doesn't work half the time—too heavy a load." He also has to pick up more frames per hour and therefore has more physical work. This he regards as an important change. With the changed equipment two men do the work of three, but nobody was laid off in his section. He himself has never been unemployed.

He is less satisfied now with his job than he was 5 years ago, and he reported that he makes less money (presumably because he works less overtime). "You got to make sure you are not too slow and can keep up. You have to watch it closer, watch three or four pieces at a time. You have to watch for the other guy so that he doesn't lose his hands." He feels that "automation causes problems" to people in his line of work. "It takes jobs away and more are laid off." About his job: "It's steady. That's all I can say. Seniority doesn't help as far as what job they give you." He has no plans to look for another job, however.

Case 6:   **This man is an all-around helper** in a large factory which produces bowling games and pin ball machines. He is white, 58 years old, married, and has no children under age 18. He had 6 years of schooling and no further training. He says that he feels no need for further education. The interviewer reported that he is functionally illiterate.

He was born in a coal camp in West Virginia and worked for 32 years in a coal mine there. Then there no longer was enough work. Three years ago he went to Illinois and obtained a job in a typewriter factory, with such duties as stamping rubber rollers and rubber molds. The company changed a spray it used in its operations and he found he was allergic to the new spray. The company put him on another job, which he did not like. He quit. After 3 months of unemployment he found a job with a veterinarian, taking care of the animals. Although he regarded this as stop-gap work, he seems to have liked it. His pay was lowered when he took his present job in the bowling games factory. It was about the same in 1966 as 5 years earlier, only $2,000 to $3,000.

On his present job he does "a little of everything—spot welder, drill press operator. . . . I do everything in the plant . . . just an all-around man. . .run the machine, watch over it. . .I work on the line." He uses an automatic punch press and a spot welder. Although this is quite different from any equipment he was accustomed to, according to him, the work took only one day to learn. "It's like a play job; it's so easy. I never had such an easy job. . . . Getting used to it was easy, real easy." He is more satisfied with his work than 5 years ago. "It's the bosses. I like them all. It's pleasant to work with them."

Still, the job has several shortcomings: "They don't permit any talking. I talk anyway. I like to talk a lot." The job is more closely supervised than his previous work and is less steady. At the time of the interview he was working 4 days a week. "They get all caught up on production. You only work a short week." He has had several spells of unemployment in the past 5 years, including some that lasted more than a month. He feels that automation is bad for people like himself. "It takes away the jobs. Pretty soon there won't be any jobs left."

> **Case 7:** **A skilled die maker** in an automobile plant. Thirty-six years old, white, male, married with four children. He has 12 years of schooling. Afterwards he took a night school apprenticeship program in model making and die design, which took 4 years. For the past 2 years he has been taking a correspondence course in general law. He has been with the same employer for the past 16 years, has never been unemployed, and earns $10,000 to $15,000 a year on his job.

His job involves making wood models of all the automobile body parts. For this purpose he uses drill presses, drill motors, routers, planers, jointers, and contour machines, run for the most part by air pressure or electricity. The contour machine follows a pattern automatically, the other equipment is hand-guided. He describes his equipment as "a friend," but adds this qualification: "In a sense it makes the job easier. A lot of equipment is very dangerous, and you can't afford to relax because of the danger."

Five years ago a numerical control machine was introduced. This machine takes a piece of wood and cuts out a rough pattern of the model; it all but finishes the wooden models. "It eliminates a lot of our work. When this machine is perfected, it will make the wooden model from start to finish. This will eventually eliminate my job." At present the numerical control machine does most of the preliminary work on the model. He and his co-workers do the finishing work. There is less hand work, and fewer people are needed to do the job.

The change-over to the new machine was planned by the employer alone, and (according to this respondent) the men were given no notice at all. He complains: "The people who will be displaced by this machine will reach the age where their earning power decreases, and they have no chance to be involved in another type of work." About 25 percent of the workers in his section were laid off and unemployment averaged about 5 months. He himself was not laid off but, his earnings were reduced since he now works less over-time. Overall his pay is about the same as 5 years ago, and he expects no increases. He feels that the chances for advancement are poor "because the people that are used on new jobs are hired off the street. I am not being given a chance for any new openings that become available." He needed no training to work with the new machine.

His job now satisfies him less than 5 years ago. "The biggest part of my work was eliminated. I am not able to produce a decent day's work...I felt cheated in a sense of not being able to accomplish things from start to finish with my hands. . . . The machine has taken away the creativity of my job. I just put the finishing touches on what it does." The physical effort involved in his work was reduced by the new machine, but the speed with which he has to work was increased.

This man feels that automation "Is not good for the workers, but good for the company. Production costs are cut and less time is needed." He feels that the unions should take care that the workers concerned are involved in planning for technological change and that they receive any additional training they need. When asked whether additional education or training might be use-ful to him, he replied "Yes, because other types of job training I have are being outmoded."

He has some vague idea that he might like to change jobs. The disadvan-tage would be possible loss of job protection, the advantage "more money."

Case 8: **A woman riveter** in a large machinery company, white, 24 years old, separated from her husband, one child. She had 12 years of schooling but no further training of any kind, feels that she needs no further education or training.

She used to be a cashier in a grocery store where she used the cash register, a meat slicer, and also manned the switchboard. Before the birth of her child she left the grocery store and stayed out for 5 months.

After the child was born she took the riveting job, which pays more. She uses a manually controlled riveter "I hold it in place and rivet two parts together. I do the same thing all day. I just rivet parts together." The job can be learned within a day's time, she says. Speaking of her equipment, "I couldn't call it a friend. I guess I'd call it a foe because it's so monotonous."

In most respects she views the new job unfavorably compared with the previous one. The work is paced by the production line. It requires more speed and a greater physical effort. The physical surroundings are less pleasant. "Occasionally I talk.   You kind of have to talk loud because it's so noisy." She feels the work is more dangerous and more monotonous. Chances for advancement are poorer. "In the store there were jobs that paid more that you could work up to; but at. . .you just stay where you are." She also complained about having to get up earlier. For all these reasons she is less satisfied with her present job.

This woman is paid on a piece rate basis and made $5,000 to $6,000 in 1966, substantially more than 5 years ago. She has been laid off several times during the past 5 years, including at least one spell of more than a month.

In her opinion, automation "causes problems. . . . because advances in machinery eliminate jobs and they cause prices to go up because someone has to pay for it." She thinks the union should protect jobs. Some days her job feels like drudgery. "To be truthful with you, I hate it, but I need the money. To begin with, I think they should cut out piece work. I do the same thing about 1,000 times a day."

> Case 9:   **A 49-year old woman,**   white, who grew. up in rural Kentucky. Her husband is disabled and has not worked for the past 10 years. She had 8 years of school, no vocational courses, and sees no need for further education. Her income of $7,500 to $10,000 is 15 percent lower than it was 5 years earlier. She has never been unemployed.

She now works on an assembly line where she assembles and tests transmitters for telephone receivers, a job, she says, it takes at least 3 months to learn. She has been with this large telephone manufacturing company for 11 years; 3 years ago she was transferred to the current job.

Previously she used to wind coils, but the company moved the equipment to an area with lower labor costs and installed new, more automatic coil winders. The company alone planned the change. The employees had a week's notice, which she felt was sufficient. A lot fewer people were needed to man the new machines. None of the workers became unemployed because "they always give them something else to do in the plant." Her own job was abolished.

She was transferred at no change in pay or seniority rights to her current job on the assembly line after 40 hours of on-the-job training. She dislikes the new job. It requires more physical work and speed, but less skill than her former work. She claims the work is less steady now and she has less chance for advancement because, as she said in her main complaint about the transfer, "everyone has more seniority than I have."

To her, the work is more monotonous, because "in assembly, you don't have to think. I like to keep occupied." She calls the assembly line a foe, "You have to struggle to keep up." From her own experience automation causes problems because "it keeps changing people around, putting them on jobs they don't like. This used to be a better place to work."

Though less satisfied with her job, she plans no change.

Case 10: **A laborer on the river,** Negro, male, 39 years old and married with 7 children, has lived all his life in small towns in Louisiana. He has had 6 years of schooling and has taken two courses in a local trade school, paid for by the Veterans Administration. The first was a course in carpentry, which he did not finish. The second was a one-year course in auto mechanics (5 hours a day), which he did complete. He feels no need for additional education: "You see I want my children to get more education. I don't need any more. "His income is below $2,000 a year.

For the last 17 years this man has worked on a river boat with a sinking unit. "I push a button and a slab falls. It falls into the river along the river banks to keep the river banks from caving in. We make concrete slabs for the river bank. . . . We sleep on the boat. We stay for periods of 2 to 3 months and then come out for the rest of the year. I am a follow-up for a man who operates the machine that makes the slabs. And as soon as it comes down, all I do is punch the button and the slab breaks by air compression and falls. So I depend on him, how fast he operates the machine tells me how fast to go."

Formerly the slabs had to be pulled by manpower. Then new attachments were put on the machinery so that the slabs could be handled by push button. "It makes my work easier, but faster too. There is less hand work. You don't have to strain so much. I do more work—a lot more work—and it's a little easier." The same number of people are needed on the boat, but the work is completed faster. "We just finished the job sooner and we were laid off earlier. Last year we worked a full 3 months and a few weeks, this year 2 months and 2 weeks." However, the pay rate was raised. The employer planned the change-over alone and gave no notice. No training was needed to work with the new machine. "They just showed me how to work the button, and I started doing it."

His attitude toward automation is indifferent. He has mixed feelings about his job: "I like punching that button." But, on the other hand," it's kind of hard and steady with those long 10-hour schedules." He may change jobs: "I like the pay scales, and I might have to keep it. The shortened work months might make me change." He was seasonally unemployed at the time of interview and had been for 5 to 6 months. ". . .nothing else to do around here now." He says he definitely has no plans to move.

Case 11:   A 57-year old white man, self-employed.   Had 12
years of school and 3 semesters of college; no vocational training,
never unemployed; 1966 income of $7,500 to $10,000 was the
same as 5 years earlier.

This man has owned and operated an automobile service station for 3½
years. Previously he owned an automatic transmission business. He and his
employees use a power drill, pneumatic hammer, and motor analyzer to repair
cars. He himself uses a newly acquired chassis machine, which straightens the
chassis. It saves much hand labor and time: "Before we had to straighten the
chassis by hand and pulleys. Now it's done by hydraulic and pneumatic
power."

In order to use the machine, he had to learn leveling and synchronizing
of torsion bars, alignment, and motor analysis. He says that he did this just by
reading: "a mechanic knows what he is reading—how to interpret it." He
believes that it would take someone else "at least 3 years" to learn to work
with the equipment in this station plus at least 10 years of background in the
automotive business.

The new equipment requires less physical work and skill and makes his
job less interesting: "It's not challenging anymore. Now it's cut and dried. The
machine operates for you. No special handskill is used." He needs fewer
employees now to do the same amount of work.

His fondness for the "experienced old mechanic equipped with mental
knowledge and hand tools" is evident in his criticism of new workers: "The
old-time men retire and the new school of men come into this field. Although
they come from vocational automotive schools, they are only book-learned
and have no basic fundamental experience in the actual process of the work
or reading of diagnostics pertaining to automotive repair."

As might be expected, he regards his equipment as a foe, and complains
that automation "takes away the skill from industry."

He enjoys his work, but is less satisfied than 5 years ago, and probably
will sell his business for economic reasons: "With prices unchanged, and taxes
and commodities and salaries increasing, business is a hard and torturous proposition."

Case 12:   A 32-year old white woman, married with 5 children;
husband a brick layer for construction jobs. She grew up on a
farm in Kentucky, had 12 years of school plus vocational courses
in typing and bookkeeping. She would like to use these skills. Her
income of $2,000 to $3,000 has increased $14 a week over the
last 5 years. She has had two or three spells of unemployment
plus time out for a child's arrival in the last 5 years.

This woman has had a full time job with a large tobacco company for the past 2 years, and she also worked there previously. "I run a machine called the cellophane and banding machine. I place cigars in the machine, and put cellophane and bands in the machine. The machine puts the bands and cellophane on and seals them. Next I catch the cigars and put 50 in a box." She likes operating this machine. Her biggest complaint about her job centers around another machine connected to the first one which she also operates, called a 5-pack machine, which puts five cigars in a box. "It's hard for me to adjust to. The boxes are difficult for me to put in the machine and the hopper for cigars is fed by hand."

About 6 months ago the company installed a new type of cellophane and banding machine that works twice as fast. As a result speed and amount of physical work are greater. "I feed twice as many cigars into the machine and I catch twice as many wrapped cigars to put in boxes." No one became unemployed because the company doubled its output.

Her negative attitude toward automation and machinery change is a direct result of her unpleasant experience, especially as it involved the 5-pack machine. "It is hard to feed the hopper for the 5-pack machine. The new machine has to be fed twice as fast and 5-pack cigar boxes also have to be put in faster. . . . There should be automatic hopper fillers for the 5-pack machine. Loading is done by girls and is too hard on them."

She regards the job as drudgery because of "that machine, night work, and 3 hours of commuting time each day." She might change jobs to be closer to home and have more time with her children.

She recognizes the need for more education: "I should have gone to college. I could have a better job with more education." But it would be impractical because "I need to be with my five children, and of course I don't have finances for more education now."

*   *   *

The reader may be struck by the fact that despite the rather comprehensive selection criteria, most of the dozen cases identified as having made a particularly poor adjustment to technological change are not genuine hardship cases. It appears once again that in the expanding economy of the mid-1960's the *direct* impact of technological change seldom was such as to cause serious distress. There were of course other workers in the sample who suffered a good deal of unemployment, income declines, or low earning levels; but their plight cannot be linked *directly* to technological change. They experienced no change in machine technology during the past 5 years.

Chapter 8

THE ROLE OF EDUCATION IN RELATION
TO TECHNOLOGICAL CHANGE

Do advances in machine technology demand an increasingly educated labor force, or do they lower educational and skill requirements? It is often argued that job openings which are available in technologically advanced industries call primarily for educated and skilled workers, and that the educational system must be geared up to meet the challenge of technological change.[1] At the same time the opposite idea also finds frequent expression— that highly mechanized or automated jobs give workers only a minimal opportunity to use education and skills.[2] Considerable empirical evidence has been marshalled to support each of the opposing views.[3] Much of this conflicting evidence consists of case studies of particular plants or industries which have made rather drastic changes in machine technology. It is quite possible that in some cases technological advance requires on balance more educated workers, and that the opposite is true in other cases. A series of case studies conducted by the Manpower Research Unit of the British Ministry of Labour concluded:

---

[1] A good exposition of this view is the testimony by Charles C. Killingsworth before the Senate Subcommittee on Employment and Manpower in the fall of 1963; appearing in the Committee's *Nation's Manpower Revolution,* Part 5, U. S. Government Printing Office, 1964.

[2] See for example James R. Bright, "The Relationship of Increasing Automation and Skill Requirements," *Report of the National Commission on Technology, Automation and Economic Progress,* Appendix Vol. II, pp. 207-221; also Edwin L. Dale, Jr., "The Great Unemployment Fallacy," *The New Republic,* September 1964.

[3] For a summary of this literature see Morris A. Horowitz and Irwin L. Herrnstadt, "Changes in the Skill Requirements of Occupations in Selected Industries," *Report of the National Commission on Technology, Automation and Economic Progress,* op. cit., pp. 227-230.

"Within a broadly defined skill group such as skilled operatives, in some cases new technology 'de-skills' operations, while in
others it increases the skill required. The average effect on the
whole group is a resultant of a great many movements within it,
many of which will be in opposite directions."[4]

A thorough U.S. study of five industries which experienced striking advances
in mechanization and automation over the past 15 years came to a similar
conclusion:

"The overall or net change in the skill requirements of
occupations in these industries was remarkably small, despite the
15 years covered. One industry on balance had an increase, one a
possible decline, but in each case the shift was modest. Moreover,
substantial changes in occupational content were not common, and
the number of obsolete occupations was few. However the small
net change in skill levels was the product of numerous offsetting
changes in the various abilities needed for individual occupations
in an industry.   There was considerable change in occupational
requirements and content, but on balance it was either inconsequential or inconclusive with respect to overall skill level."[5]

Case studies cannot quantify these contrary tendencies. The survey permits us
to measure the way in which various levels of technology are related to the
education and training of the work force, taking account of both formal
education and vocational training (described in Section A of this chapter).[6]

It is evident that a worker's education and training are not always
ideally suited to his job. In a sense hiring practices are adjusted to the available worker qualifications. Therefore we shall not be satisfied with describing
the level of education which merely happens to be associated with various
levels of machine technology. In Sections B and C following we shall raise a
further question: To what extent do present levels of education represent
excessive or insufficient amounts of education and training in relation to job
demands? This difficult question might be answered differently by educators,
personnel experts, labor economists, engineers, or union officials. The survey

---

[4]Sir Denis Barnes, "Technological Change and the Occupational Structure," Pro-
ceedings of the International Conference on Automation, Full Employment, and a
Balanced Economy," Rome, 1967, The American Foundation on Automation and
Employment, Inc., p. 5.

[5]Horowitz and Herrnstadt, *op. cit.,* p. 287. The study is based on the changing
descriptions of job content appearing in dictionaries of occupational titles over the past
15 years.

[6]In this chapter the expression "formal education" refers to years of schooling,
including college, as distinguished from vocational and other special training, which is
measured separately.

merely gives some indication of how members of the labor force view their own education and training in relation to their work, and how education may influence attitudes toward the use of machines.

Finally Section D attempts to explore the diverse ways in which education may facilitate adjustment to technological change. In so doing we shall distinguish between three kinds of job prerequisites: Formal education, vocational skills, and what for want of a better term are called "personal capabilities." The latter are the capabilities which enable a person to respond to the kinds of demands measured by the Job Demands Scale: to adjust to a fast work pace, to be willing to learn and to move from one specialty to another, to take responsibility, to plan, take initiative and use judgment in a setting where the work flow is becoming increasingly complex and different operations are becoming increasingly integrated. If the educational system is to prepare people to work in this kind of setting, more should be known about the kinds of qualifications which would be most useful to different groups in the labor force. While some data are presented in this last section, it serves primarily to point up the need for further research.

A. Level Of Machine Technology And Education

It was shown in Chapter 2 (Table 2-4) that the formal education which members of the labor force possess is not decisive in determining whether they will work with machinery on their job. In every major education group at least two-thirds of people use machines on their job either directly or indirectly. College graduates, along with those who have only a grade school education, are relatively numerous in the group which does not work with machines at all. College graduates also are found with relatively high frequency in the group which has only indirect contact with machines. Yet differences in machine use between educational groups are moderate. An important finding is that about 45 percent of college graduates and about 65 percent of those with a high school degree *operate* machines at least occasionally on their jobs.

Since machine use is common in all educational groups, one might expect that all educational groups would experience machine change with similar frequency. Data presented in Chapter 3 indicate however that those with a high school degree or more formal education experienced more technological change in the past 5 years than others who remained in the same job (Table 3-7). This finding suggests that the better educated are more likely than those with less formal education to move up to jobs on which they use the newer and more sophisticated kinds of equipment. Table 8-1 corroborates this inference. It reveals a striking relationship between formal education and

TABLE 8-1

DISTRIBUTION OF WORKERS BY THE AUTOMATION LEVEL
OF THEIR EQUIPMENT AND EDUCATION

Automation level of equipment operated directly	Education						No. of cases
	0-7 grades	8-11 grades	High school degree	College[a]	Not ascertained	All	
Numerical, tape, computer, or other logical control	1%	6%	24%	69%	*	100%	106
Fixed mechanical control	5	25	35	33	2%	100%	1,038
Powered multi-system, manual main control	16	40	31	12	1	100%	409
Manual control, operator powered, or powered single-system	10	40	33	15	2	100%	348
Automation level of equipment used indirectly							
Numerical, tape, computer, or other logical control	*	6	26	67	1	100%	173
Fixed mechanical control	4	29	32	33	2	100%	405
Powered multi-system, manual main control	8	38	32	21	1	100%	157
Manual control, operator powered, or powered single-system	2	43	43	12	*	100%	42

*Less than 0.5 percent.
[a]This group includes all those with college experience of at least one year.

TABLE 8-2

FREQUENCY OF VOCATIONAL EDUCATION AMONG WORKERS
WITH DIFFERENT AMOUNTS OF FORMAL EDUCATION

Formal education	Vocational training[a]				Total	No. of cases
	None	One course	Two or more courses	Not ascertained		
0-7 grades	90%	5%	2%	3%	100%	212
8-11 grades	64	25	8	3	100%	763
High school degree	29	50	19	2	100%	854
Some college[b]	38	44	16	2	100%	407
College degree	56	30	11	3	100%	383
All	49	35	13	3	100%	2,662

[a]Vocational training courses of 3 months or more.
[b]One to 3 years of college including all degrees below the B.A. level.

the automation level of equipment employed by the work force. About one-half of the people who *operate* manually controlled machinery have not completed high school, while a mere 7 percent of those who operate the most automated categories are at this low educational level. Conversely we find only 15 percent college educated among the group at the lowest level of the Automation Scale, compared with nearly 70 percent at the upper end. For people who have indirect contact with equipment, the relationship between education and degree of automation of that equipment is nearly as pronounced.

One might assume that vocational education and special training programs would be as relevant as formal education to the kind of equipment which a worker uses. In the survey all members of the labor force were questioned regarding a broad range of vocational courses or special training programs which they might have taken in high school, in the Army, in vocational schools, on the job, under union auspices, in evening classes, by correspondence courses, and the like. Somewhat arbitrarily, courses or programs which lasted less than 3 months will be disregarded in our analysis on the grounds that they are unlikely to raise the workers' occupational qualifications materially.[7] Almost exactly one-half of the labor force reported that at some time in the past they went through a vocational or special training program thus delimited.[8] Vocational training was reported by a higher percentage of people who operate machines on their job (58 percent) than by those who do not operate any equipment (42 percent). Most of the people who have had some occupational training took only a single course or program. However, 10 percent have had two, and 3 percent three or more, such training experiences. In terms of absolute numbers more than 10 million members of the labor force have had two or more vocational or special training programs lasting 3 months or more.

Table 8-2 shows that vocational and special education are highly related to formal education. The workers who in the sense have the greatest need for vocational skills are least likely to obtain them, whether by choice or circumstances. Less than one in ten of workers with 7 years of schooling or less have any significant occupational training. The proportion with at least 3 months of occupational training is well below four in ten for people with 8 to 11 years of school completed and jumps to seven in ten for high school

---

[7]Many of these courses or programs met only a few hours a week.

[8]This estimate agrees closely with a 1963 estimate derived from a survey conducted by the U. S. Bureau of the Census for the U.S. Department of Labor, *Formal Occupational Training of Adult Workers,* Manpower/Automation Research Monograph No. 2, although there are several minor differences in coverage and definitions between the two surveys.

graduates. It declines to six in ten among those with 1 to 3 years of college and to about four in ten for college graduates. Needless to say, the kind of training received also varies with formal education. A 1963 study by the Census Bureau and the Bureau of Labor Statistics confirms the close association between occupational training and years of schooling. That study comments:[9]

> "Training is closely linked with education, largely because of the important role of high schools and other educational institutions in developing job skills. Moreover, high school graduation is now typically required for admission to industry training programs, both company training schools and apprenticeship. In the military services, too, which are an important source of training, men with more education tend to score better on the aptitude tests used for selecting and training personnel."

We may now turn to the relation between vocational or special training and advances in machine technology. Chapter 6 showed that about 55 percent of those who experienced a change in machine technology felt that more skill was needed after the change (Table 6-6). One might suppose that vocational education and special training programs would be instrumental in preparing people to acquire the needed skills. This supposition seems to be only partially correct. About 62 percent of those who work with computers or other logically controlled equipment have had some vocational or special training lasting at least 3 months. For workers who operate manually controlled equipment this proportion ranges between 40 and 50 percent. Moderate differences in vocational training also appear between workers classified according to the automation level of equipment with which they have indirect contact. The important finding is that people using equipment at successive levels of mechanization and automation differ less with respect to the incidence of vocational training than with respect to formal education. Indeed the differences in vocational education in Table 8-3 may reflect largely differences in formal education (with which vocational education is so strongly associated). Occupational training, unlike conventional schooling, may be obtained as a consequence of an equipment change. The evidence that, despite the existence of training programs which prepare workers for the use of new and more complex machines, occupational training rises relatively little with the level of mechanization and automation is suggestive. It would appear that much of the training needed to work with new equipment is acquired on the job by informal means such as help from the supervisor or fellow workers, and learning-by-doing. This inference is consistent with data presented in Chapter 4. It

---

[9]*Ibid.,* p. 5.

TABLE 8-3

VOCATIONAL TRAINING BY THE AUTOMATION LEVEL OF WORKERS' EQUIPMENT

Automation level of equipment operated directly	Vocational training[a]				
	None	One course	Two or more courses	Not ascertained	Total
Numerical, tape, computer, or other logical control	35%	45%	17%	3%	100%
Fixed mechanical control	42	41	14	3	100%
Powered multi-system, manual main control	56	31	10	3	100%
Manual control, operator powered, or powered single-system	50	33	15	2	100%
Automation level of equipment used indirectly					
Numerical, tape, computer, or other logical control	37	41	20	2	100%
Fixed mechanical control	41	36	19	1	100%
Powered multi-system, manual main control	51	29	16	4	100%
Manual control, operator powered, or powered single-system	43	38	17	2	100%

[a]Vocational training course of 3 months or more.

may be recalled that only a small proportion of workers who experienced a machine change during the past 5 years under study reported that they received any kind of organized training during the transition (Table 4-2). Many more said that they merely received some on-the-job training or taught themselves; and a large group said no training at all was needed to work with the new equipment.

One gets the impression from the survey data that the kinds of demands the advancing technology makes on workers often are not ones that are most readily met by organized occupational training programs. On the one hand, very specialized knowledge of particular machine operations or work tasks often seems to be taught on the job. On the other hand, we saw in Chapter 6 that technological change demands in addition some personal capabilities—versatility, adaptability, speed, alertness, judgment, initiative, and interest in learning. Little is known about the means by which such personal capabilities can be developed or enhanced. Formal education and vocational education may play some role together with experience, childhood training, and inborn traits. Also, these capabilities are not readily measurable. In the absence of relevant information, employers who match workers and job openings may see a relatively high level of schooling as indicative of the presence of these personal capabilities.

Table 8-4 gives an indication of the kinds of occupational training associated with various levels of machine technology. The data are shown separately for blue-collar and white-collar workers. The top line of each part of the table shows the proportion of the group who have had occupational training lasting 3 months or more. As noted earlier, this proportion increases, but not a great deal, with automation level of equipment. Further, the table presents a distribution of *courses* by general subject matter for workers at the various levels of machine technology. Since one person may have taken several courses, a distribution of courses is different from a distribution of people. Vocational courses taken in school are combined with those taken later. We find, of course, that white-collar workers have taken commercial, business, and semiprofessional courses with greater frequency than blue-collar workers; blue-collar workers have a higher proportion of courses in auto mechanics, machine shop, and other mechanical subjects as well as carpentry, electronics, and electrical work. What is most interesting, however, is that many white-collar workers have had courses which prepare a person for blue-collar work and vice versa. Moreover, the kinds of courses taken by people working with equipment of varying degrees of automation do not show a distinctly different pattern. The data give the impression (although they do not prove this) that much occupational training must be unrelated to people's present work. A look at individual interviews strengthens this impression. People who have gone through several vocational courses or training programs often have acquired some rather diverse skills: for example, radio and electronics in the Army and later a vocational course in carpentry; or to give another example, a company training program in modern accounting practices plus a high school course in auto mechanics and an evening school course in interior decorating.

B. Education And Attitudes Toward Machine Technology

At the attitudinal level, the relation between education and the use of machines on the job may be investigated by comparing attitudes towards machines and automation among workers at different educational levels. Does the use of equipment enhance the interest of the job for the educated worker, or does the well-educated worker resent the growing intrusion of machines into every job and occupation?

TABLE 8-4 (Sheet 1 of 2)

DISTRIBUTION OF COURSES TAKEN BY WHITE-COLLAR AND BLUE-COLLAR WORKERS OPERATING VARIOUS TYPES OF EQUIPMENT

	All	Numerical, tape, computer or other logical control	Fixed mechanical control	Powered multi-system, manual main control	Manual control, operator powered, or powered single-system	No equipment
			Blue-collar workers			
Proportion of workers with vocational training	47%	a	51%	43%	52%	41%
Type of training						
Agriculture	8%	a	4%	18%	4%	6%
Woodworking, carpentry	5	a	6	2	5	4
Metalworking, auto and other mechanics	35	a	36	38	40	26
Mechanical drawing, drafting	4	a	2	4	3	6
Electricity, electronics, radio, television	12	a	8	14	20	10
Semi-professional	11	a	16	5	9	13
Commerical, business	10	a	14	6	8	13
Other	12	a	11	9	11	20
Not ascertained	3	a	3	4	*	2
Total	100%		100%	100%	100%	100%
Number of cases	(951)		(335)	(212)	(206)	(191)

*Less than 0.5 percent.
aThere are insufficient numbers of blue-collar workers with this type of equipment for meaningful analysis.

TABLE 8-4 (Sheet 2 of 2)

DISTRIBUTION OF COURSES TAKEN BY WHITE-COLLAR AND BLUE-COLLAR WORKERS
OPERATING VARIOUS TYPES OF EQUIPMENT

	All	Numerical, tape, computer or other logical control	Fixed mechanical control	Powered multi-system, manual main control	Manual control, operator powered, or powered single-system	No equipment
			White-collar workers			
Proportion of workers with vocational training	57%	67%	64%	45%	56%	47%
Type of training						
Agriculture	2%	2%	1%	7%	2%	4%
Woodworking, carpentry	3	3	2	7	3	5
Metalworking, auto and other mechanics	16	23	11	28	34	20
Mechanical drawing, drafting	5	7	3	*	5	10
Electricity, electronics, radio, television	7	5	4	7	10	11
Semi-professional	10	5	13	*	8	8
Commercial, business	37	27	50	16	21	23
Other	17	24	15	30	15	15
Not ascertained	3	4	1	5	2	4
Total	100%	100%	100%	100%	100%	100%
Number of cases	(1,042)	(110)	(545)	(43)	(59)	(285)

*Less than 0.5 percent.

TABLE 8-5

EDUCATION AND WORKERS' ATTITUDES TOWARD THEIR JOB AND EQUIPMENT

	Formal education				
	0-7 grades	8-11 grades	High school degree	Some college[a]	College degree
Job is					
Enjoyable	71%	70%	78%	82%	89%
Pro-con	22	22	16	14	8
Drudgery	5	7	4	3	2
Not ascertained	2	1	2	1	1
Total	100%	100%	100%	100%	100%
Number of cases	(212)	(763)	(854)	(407)	(383)
Automation is					
Good	19%	25%	36%	43%	47%
Pro-con	3	3	3	3	3
Bad	14	15	11	8	4
Makes no difference	59	55	48	43	42
Not ascertained	5	2	2	3	4
Total	100%	100%	100%	100%	100%
Number of cases	(212)	(763)	(854)	(407)	(383)
Attitude toward equipment[b]					
A friend	70%	70%	72%	77%	73%
Pro-con	15	16	15	10	16
A foe	5	6	5	6	2
Not ascertained	10	8	8	7	9
Total	100%	100%	100%	100%	100%
Number of cases	(137)	(544)	(617)	(270)	(216)

[a]One to three years of college including all degrees below the B.A. level.
[b]Asked only of those who use equipment.

By way of background, the satisfaction which people derive from their job seems to rise to a moderate extent with education. Answers to the question whether people enjoy their work or whether it is drudgery illustrate this relation (Table 8-5). Whereas college graduates say in nearly 90 percent of cases, that they enjoy their work and seldom qualify this statement, the same is true of only about 70 percent in the lower educational categories. The less educated often gave pro-con answers—in some ways they enjoy their job, in others they dislike it.

Much more pronounced differences emerge in replies to the question: "In general would you say that automation is a good thing for people doing your kind of work, or does it cause problems, or it doesn't make any difference?" Attitudes toward automation grow much more favorable with increases in formal education. Educated workers not only use more automated equipment than others; in the large majority of cases they seem to like this development. To those with less education automation is more often threatening, probably because the less skilled and less complex jobs of those people are more easily taken over by machines than those of more highly educated members of the labor force. Even in the lower education groups, however, the majority hold that automation makes no difference or is a good thing. Clearly there is no evidence that people with a good deal of formal education feel that the increasing use of machines in connection with their work and the increasing capability of the machines they work with are incompatible with their education.

The introduction of modern equipment may add to the scope and variety of the educated person's work, hence his job satisfaction and favorable attitudes toward automation. There is no indication, however, that the greater enjoyment of the job among the better educated groups derives very directly from the equipment they work with. Expressed attitudes toward the particular machine operated by the respondent are similarly favorable at all educational levels. If anything, there is a greater disposition in the higher educational categories to call the machine "a friend" or "a helper;" but those differences are barely perceptible.

C. Felt Need for Education

To make a meaningful investigation of the appropriateness of workers' education in relation to various levels of mechanization or automation would require extensive surveys devoted wholly to that purpose. The analysis which follows has a much more limited scope: we want to know whether the high levels of education associated with technologically sophisticated equipment are felt to be excessive by the labor force members concerned. Or contrariwise, do people who work in a technologically dynamic setting have a felt need for additional education? Two measures are used to explore these questions. First, all members of the labor force were asked, "In connection with your future work do you feel that it would be useful for you to get additional education or some kind of training, or is there no need for it? (*If additional education wanted*) Do you think that you might get such education or training in the next few years, or is the idea impractical?" A probe asking for reasons followed both questions. Secondly, immediately following the questions on vocational and special training, all members of the labor force were asked, "Through your previous experience and training, have you built up some skills that you would like to be using, but can't on your present job? What are they?"

In all, 44 percent of people expressed a felt need for additional education to help them with their future work, while only 17 percent said that they had training and skills which they were unable to use. Both figures require some explanation and qualification. Some of the people who said that they would like more education stated in reply to the follow-up question that it would be impractical for them actually to acquire it, most often for reasons of time, money, or advanced age. When this group is excluded there still remain 33 percent of all labor force members who seemed to react favorably to the idea of getting more education in the next few years. The major reason for wanting more education is not necessarily to do one's current job more competently, although many people gave this explanation. Since doing one's current job more competently may lead to a raise or a promotion, the people who explained that they wanted more education for the sake of performing their present work better may have basically the same motivation as many others who said, "I want to make more money" or "I want to get ahead."

The question about unused skills was asked of all members of the labor force, irrespective of whether they had ever had any occupational training. In all, 17 percent reported unused skills. The answers were related to the number of employers which the respondent had in the past 5 years and to the number of vocational courses (taken any time). About 10 percent of those who had no occupational or special training said they had unused skills; these were skills which they had acquired by experience or on-the-job training in previous employment. Included also are some people who acquired a knowledge of farming in their youth. Among those who had one vocational course 23 percent reported unused skills, and among those with two or more courses 26 percent. The 1963 study conducted by Census together with the Bureau of Labor Statistics found a considerably higher frequency of unused vocational training. In all, 51 percent of courses taken by male and 42 percent of those taken by female members of the labor force were reported by the Census-BLS to be unused on the present job, although one-half of these unused skills had been of value on a prior job.[10] The major reason for the lower figure obtained in

---

[10]*Op. cit.,* p. 40. The Census-BLS study measured unused training programs or courses, whereas the present survey measured the presence of any unused skills. People who have taken several courses and use some but not others, would appear in the BLS

the present study lies in the wording of the question. The question asked in the survey here referred to unused skills, *"which you would like to be using,"* while the Census-BLS questions referred to unused skills in general. Many people have advanced during their working lives and are happy *not* to be in the jobs for which certain vocational courses might have once prepared them.

Even the low overall estimate of 17 percent with skills which they would like to use is in some sense an overstatement; for there was occasional wishful thinking on the part of respondents. For example, a man who feels that he knows how to paint would like to be a book illustrator. One respondent majored in education in college and then went into his father's business, where he earns at least twice as much as a teacher; his wish to use his educational training exists side-by-side with an unwillingness to sacrifice his higher business income. However, such cases as these constitute a minority.

Age and education have some influence on perceived need for additional education and on perceived use of skill and training. The felt need for additional training is strongly associated with being young; and it rises sharply with the level of formal education already attained. Less expected perhaps is the finding that young and well-educated members of the labor force also have a somewhat greater tendency than others to see themselves as having unused training and skills, possibly an indication that their present job falls short of their aspirations.

Technological change and job change contribute importantly to the felt need for additional training as well as to the existence of unused skills. In Table 8-6 the labor force is divided into four groups: people without job change or technological change in the past 5 years, those experiencing a change in machine technology while remaining in the same department of the same company, those having one or more job changes or transfers but no change in the equipment they work with, and those having both a job and an equipment change. It is apparent from Table 8-6 that technological change significantly raises the felt need for additional training, although not quite to the degree that job changes do. Among those who had both job and equipment changes, as many as 46 percent expressed an interest in more education and training (the first row in Table 8-6 *excludes* workers who said they would like more education and training but thought it would be impractical to get it). The proportion seeing themselves as having unused skills is raised to a lesser extent by technological change and also by job change. The two variables show the same response to job and equipment change, in part no doubt because finding that old skills have become useless may stimulate a desire for

---

(Footnote 10 Continued)

figures both under the used and the unused category, in the present survey only under the unused category. The low estimate in this survey of 17 percent with unused skills *which they would like to use on their job* is all the more noteworthy.

TABLE 8-6

PERCEIVED NEED FOR ADDITIONAL EDUCATION AND PERCEIVED USE OF SKILLS
AND TRAINING RELATED TO MACHINE CHANGE AND JOB CHANGE

	Same job for past 5 years		Different job[a]	
	No machine change	Machine change	No machine change	Machine change
Proportion of workers who feel that additional education would be useful[b]	26%	35%	42%	46%
Proportion of workers who feel that they have unused skills	13%	17%	20%	25%
Number of cases	(1,165)	(170)	(642)	(363)

[a]Includes those who were transferred within the same company.

[b]Includes only those who also feel that such education would be practicable to get.

new training and skills.[11] However, in every group about twice as many people had a felt need for more education or training as reported unused skills. The figures reflect a tendency of the U.S. labor force to be education-oriented, in the sense that education is seen as a path toward advancement and status. The point to be emphasized in the context of this monograph is that changes in machine technology are associated with greater perceived needs for education than technologically static job situations.

Table 8-7 leads to similar conclusions. It presents the frequency of felt need for additional education as well as the frequency of reporting unused skills by automation level of equipment. It is quite clear that at every automation level, and also among people who do not work with equipment, more people expressed an interest in additional education and training than said that they had unused or obsolete skills. The felt need for education rises somewhat with automation level of equipment; however, by far the largest difference is that between people who work with computers and other logically controlled equipment as against all others. In the highest automation category more than half of all workers expressed an active interest in additional education and training. The presence of unused skills appears to be unrelated to automation level, although people who do not operate any equipment are somewhat less likely than others to report unused skills.

---

[11]In the population as a whole those who have unused skills expressed a felt need for additional education or training more frequently than other members of the labor force.

TABLE 8-7

AUTOMATION LEVEL RELATED TO PERCEIVED NEED FOR ADDITIONAL EDUCATION,
AND PERCEIVED USE OF SKILLS AND TRAINING

	Proportion of workers who feel that additional education would be useful	Proportion of workers who feel that additional education would be both useful and practicable to get	Proportion of workers who feel that they have unused skills	No. of cases
All	43%	33%	17%	(2,662)
Automation level of equipment operated directly				
Numerical, tape, computer, or other logical control	67%	60%	21%	(106)
Fixed mechanical control	46	36	19	(1,038)
Powered multi-system, manual main control	37	26	16	(409)
Manual control, operator powered, or powered single-system	41	32	21	(348)
No equipment operated	42	30	12	(743)
Automation level of equipment used indirectly				
Numerical, tape, computer, or other logical control	64	53	14	(173)
Fixed mechanical control	50	36	19	(405)
Powered multi-system, manual main control	48	40	17	(157)
Manual control, operator powered, or powered single-system	43	28	17	(42)
No indirect use of equipment	40	30	17	(1,857)

These findings underline the importance of education in relation to the modern technology. Not only do people who use more mechanized or automated equipment have more education than other workers; they also feel greater need for further education. To be sure, the youth and relatively high educational attainment of this group accounts in part for its interest in education. Despite their greater educational attainment, there is no indication that people who work with highly mechanized or automated equipment believe that they have unused skills to any greater extent than other groups.[12]

___

[12]A third line of inquiry neither confirms nor contradicts these conclusions. People who operate equipment were asked how much education and training is needed

D. Education And Adjustment To Technological Change

Not only is the modern technology associated with a highly educated work force; it also appears from the survey data that education normally does not complicate adjustment to more mechanized or automated equipment. Indeed education seems to facilitate adjustment to a moderate degree. The Adjustment Scale constructed in Chapter 7 is based, it may be recalled, on several measures of job satisfaction: reports regarding income change, promotions, unemployment, attitudes toward machines and automation, perceived changes in job interest, steadiness, and chances for advancement. In Chapter 7 it was shown by means of multivariate analysis that adjustment to equipment change (with or without an accompanying job change) is related to the worker's formal education (Table 7-2). People who did not complete high school had more difficulty than high school graduates and college-trained workers, and those who had less than 8 years of schooling made the poorest adjustment. The relationship persisted, though in weakened form, after allowing for the effect of other related variables such as occupation, income, and age. People with a college education seem to have made a somewhat less successful adjustment to technological change than high school graduates. The difference is small and the reason for it not clear, although the college-educated group does include a substantial proportion of people who had less than 4 years of college and also a few people who were working temporarily to earn enough money to return to college.

Vocational education may be added to the multivariate equations shown in Table 7-2. The results appear in Table 8-8 below, using Adjustment Scale A (indications of poor adjustment). The beta coefficient is very low (.05), signifying that vocational education has no appreciable influence on the kind of adjustment which people make to technological change. The corresponding

---

(Footnote 12 Continued)

for their job (*not* merely the machines they operate) and the answer was compared with the amount of education they actually have. Almost one-half of people who operate machines either mentioned only vocational courses or said outright that no formal education at all is needed for their work. Another 17 percent mentioned appreciably fewer years of schooling than they actually possess. The first group was operating manually controlled machines in relatively large numbers; the second group, on the other hand, included a slightly larger than average number of workers operating the more automated kinds of equipment. Since both groups presumably feel overeducated for their job, it is not clear what these differences imply. Moreover, many of these people expressed a desire for additional education to help them in their future work. Apparently, education is seen even by these people as a means of getting ahead or of moving out of an unsatisfactory situation. The only thing that can be inferred from this line of questioning is that respondents did not know what to make of it and how to reply; and this despite several attempts to improve the wording of the question during the pretest. Probably many workers were simply saying that their job does not involve reading, writing, or arithmetic. This variable is therefore disregarded in this chapter.

TABLE 8-8

RELATION OF VOCATIONAL EDUCATION TO ADJUSTMENT SCALE A

Mean number of indications of poor adjustment: 1.13

Vocational training[a]	Number of cases	Deviations from mean		Beta coefficient
		Unadjusted	Adjusted	
None	265	+0.13	-0.08	
One course	212	-0.07	+0.05	
Two or more courses	110	-0.18	+0.09	
				.05

[a]Vocational training courses of 3 months or more.

beta coefficient for formal education is somewhat higher (.10). It should be noted here that vocational education is of very uneven quality, probably more so than formal education. Quality is not reflected by our measurement (only the requirement that the program extended over 3 months or more). The unadjusted coefficients do show a slightly better reaction to technological change by people with some vocational education than by those without. However, the adjusted coefficients reverse this result, apparently because of the strong interrelation between vocational education on the one hand and years of formal schooling as well as occupation on the other.[13] Whether the influence of vocational education is slightly negative or positive, the important point is that in contrast to years of schooling, it does not seem to make a measurable contribution to people's ability to make a successful adjustment to technological change.

The indication that formal education facilitates adjustment to technological change, at least to some moderate extent, while vocational education apparently fails to do so, may suggest that the kinds of qualifications which the advancing technology calls for can best be met by formal education. In job situations where the newer technology demands more technical or scientific understanding, this inference is obviously valid. The intellectual skills demanded by new machines were not measured in the survey. However, case studies conducted in American firms usually have shown that technological advance creates a moderate number of new jobs with very high requirements of skill and knowledge. The case studies conducted by the Manpower Research Unit of the British Ministry of Labour show, as might be expected, increasing use of scientists and technologists. They also point to the emergence of a new class of supervisors or "supercraftsmen."

---

[13]Lack of vocational education *may* contribute toward putting a person into the occupational category—blue-collar workers, other than craftsmen and foremen. Chapter 7 indicates that this group does not adjust as readily to technological change as others.

TABLE 8-9

RELATION BETWEEN EDUCATION AND JOB DEMANDS

	Education				
	0-7 grades	8-11 grades	High school	Some college	College degree
	Job changers				
Net job demands					
Decline	0	18%	15%	10%	7%
No change	55%	38	19	18	14
1-3 increases	27	27	34	40	40
4-6 increases	18	17	32	32	39
Total	100%	100%	100%	100%	100%

	Equipment changers				
Net job demands					
Decline	28%	15%	10%	14%	10%
No change	20	15	17	18	11
1-3 increases	29	39	33	33	44
4-6 increases	23	31	40	35	35
Total	100%	100%	100%	100%	100%

"These persons in technologically advanced plants required greater mental ability in reasoning and application which greatly outweighed the physical manipulative skills possessed by ordinary skilled craftsmen."[14]

In addition, the present survey identified a complex of more general job demands which respondents perceived as rising in consequence of machine change. These included demands for more speed, more skill, the ability to plan and to exercise judgment and initiative, to organize one's own work, and the capability of taking advantage of greater opportunities to learn new things. We must ask whether more education is required when such job demands are raised. The Job Demands Scale, which measures the extent of changes in job demands which the worker reported, shows a fairly pronounced relation to education among job changers and transfers and a weaker relation among those who experienced an equipment change[15] (Table 8-9). Interest-

[14]Sir Denis Barnes, *op. cit.*, p. 5.

[15]Possibly the relationship appears weaker than it actually is because job demands are measured here in a relative sense—more or less than before the machine (and in some cases job) change.

ingly, within the very small group which made a poor adjustment there is no relation between job demands and education. Felt need for additional education rises moderately with increased job demands. The exact implication of these findings is not clear. Conceivably, the more educated workers are simply more sensitive to increased job demands. The better educated workers also may be more likely to leave old jobs in search of ones with greater demands. In addition, job demands in the sense measured may influence what kinds of workers are hired—those with more or less formal education. Whether a high level of formal education, particularly a high school degree, is really an essential prerequisite for the personal capabilities needed to meet these job demands remains a question. It has been pointed out in this connection that in Europe youngsters who leave school at the age of 14, 15, and 16 are *not* viewed as poorly fit for industrial employment and training (as they often are in the U.S.), since ages 14 to 16 is in most countries the normal school-leaving age.[16] More research in this area is clearly needed.

E. Conclusions

The survey data show first of all that the advanced technology is supported by very high levels of education among the work force using sophisticated equipment. Also, the more highly educated members of the labor force have a particular tendency to be in jobs where they experience technological change. Second, the survey data provide no indication that a significant proportion of people who work with the more mechanized and automated kinds of equipment feel that they are overeducated or overtrained for their work. There is no evidence in the survey that the amount of education which employers associate with the modern technology is viewed as excessive by the workers themselves or leads them to dislike "automation." On the contrary, a felt need for additional education is frequent, particularly among workers using the most sophisticated kinds of equipment. Third, the survey data show that the chance that a worker will adjust well to technological advance is, if anything, enhanced by education.

On the other hand, the survey was not able to clarify the precise nature of the link between educational needs and technological advance. The survey seems to indicate that formal schooling is more crucial than vocational courses for fitting workers into technologically new work settings. There is evidence, particularly in Chapter 4, that specific on-the-job training programs, rather than more comprehensive vocational courses, most frequently serve to prepare

---

[16]See Jack Stieber, "Manpower Adjustment to Automation and Technological Change in Western Europe," *Report of the National Commission on Technology, Automation, and Economic Progress,* Vol. III.

the worker for the tasks involved in his new work.[17] Formal education no doubt provides the scientific and technical know-how which may be required. Formal education should also make a worker adaptable and help him to meet the increased job demands identified in Chapter 6. The survey suggests this last idea, but the data are equally consistent with the hypothesis that employers sometimes set very high educational requirements for hiring, when these are not really a prerequisite for the work to be performed. They may do so on the supposition that completion of a good deal of formal education is indicative of the personal capabilities which *are* needed for work with technologically advanced equipment.

Detailed case studies of the work force in particular plants should be a suitable means for exploring further what specific kinds of *educational* demands are generated by the introduction of new equipment. Such studies would require interviews with workers as well as their personnel managers and supervisors. In the absence of rather extensive empirical studies of this kind it remains unclear just how the schools and the vocational training system could and actually do facilitate adaptation of the work force to continuous technological change.

---

[17]This finding is confirmed by the Census-BLS study of vocational education, *op. cit.,* p. 17.

Chapter 9

FURTHER STEPS

This study was in a sense an experiment. One of its major purposes was to find out whether a cross-sectional survey of the impact of machine technology is feasible and can yield worthwhile information. It may now be concluded that the cross-sectional approach is fruitful. It *is* possible to interview a representative sample of members of the labor force about the extent and nature of their involvement with machinery on the job, about recent changes in machinery and equipment, and about the economic and work changes which accompany technological advance. The present survey, or parts of it, could be replicated by others. Most of the questions included in it could be asked in the future of a much larger sample than was used in this first experiment. And of course there are many additional questions which suggest themselves to the reader.

Problems were encountered almost exclusively with the questions designed to reveal the technological characteristics of the equipment. It should be emphasized that the presence or absence of changes in machine technology which had a significant bearing on the respondent's job was established without particular difficulty. However, many respondents could not convey much detail about the *nature* of the technological change. Even a small, highly trained interviewing staff found the technical information difficult to obtain (and sometimes not understandable when it was obtained). Furthermore, the technical information was codable only by engineering graduates. In a more extensive study such staffing might not be available or available only at excessive cost.

There is reason to believe that the remaining data, particularly the information on economic and work changes growing out of changes in machine technology, are measured as reliably as other data obtained in carefully conducted economic surveys. Respondents were eager to talk about their experiences with machine change and replied carefully and at length to most questions. If, as we believe, respondent interest enhances the quality of survey responses, studies of people's experiences with technological advance on their job appear to be a promising field for survey research.

The subject of this study was broad in scope and dealt with a heterogeneous phenomenon—the impact of changes in machine technology on a cross–section of the labor force. The sample was of moderate size. A study of this nature can convey the major outlines of the picture but not its intimate and precise detail. It can reveal which consequences of technological change are important or unimportant, which are frequent or infrequent. It can put the often contradictory findings of case studies into their proper perspective. It can direct attention from the drastic technological changes with which case studies tend to be concerned to smaller changes, which seem to be more typical. For policy purposes the integration of pieces of detailed information is of particular importance, since many policy measures are intended to have a broad impact on a variety of situations and groups of people.

A study of this nature can also point to areas where further research might be particularly productive:

● The propositions advanced here need additional testing and confirmation through surveys which build on the experience gained in this first endeavor. Some of the survey questions might well be improved; some question sequences would be worth extending.

● The inquiry could and should be repeated with a larger sample. In the present study the number of cases in some interesting subgroups of the population is so small as to yield information only on approximate orders of magnitude. Now that the feasibility of the cross-sectional approach has been demonstrated, studies based on larger samples are called for.

● It would be interesting to replicate this survey in a period of less rapid economic expansion. In particular, the impact of advances in machine technology on employment and unemployment should be worth studying in a more stagnant economic setting.

● It would also be interesting to replicate the survey in one or two other countries. Intercultural comparisons enhance our understanding of social phenomena and of our own culture. The receptivity to change of American workers, which is so evident in the results of this survey, may be matched to varying degrees in other countries. How about the French, the English, the Spanish, the German, or the Japanese worker—how do they differ in receptivity to change and attitudes toward machines, and why?

● Instead of asking workers to report changes that have occurred on their job, one might measure change by interviewing the same workers at successive time intervals. This longitudinal or panel approach reduces problems of recall and memory distortion which may affect some of the present data. For instance, a panel of workers might be interviewed once a year over several years to trace equipment change, unemployment, income change, work changes and attitude change without reliance on memory. Panel studies could also identify more clearly than is possible with a one-time cross-section survey time sequences of events as well as the circumstances leading to and associated with a job change. The one-time approach embodied in the present study was much less costly than a panel study would have been; it also yielded answers much more quickly. Now that the feasibility of interviewing a cross-section of the labor force about experiences with technological change has been established, a longitudinal study would be valuable for validation purposes. It could also provide additional information and insights.

● Another fruitful study might be one which utilizes a sample of people who have left the labor force in a recent period to investigate the relation between advances in machine technology and withdrawals from the labor force.

● The evidence in this study that automation and mechanization raise job challenge and that job challenge in turn enhances job satisfaction should be subjected to further tests by intensive surveys with particular occupation groups as well as by case studies.

● Management practices and styles of supervision could not be studied within the scope of the present survey. Different systems of supervision may be appropriate in different technological settings; and management practices may affect the workers' adjustment to new machines. Conceivably these matters (as perceived by workers) could be studied on a cross-sectional basis, using question sequences about equipment and equipment change developed here.

● Most important perhaps would be further studies to clarify the relation between technological advance and education. The survey findings in this area raise as many questions as they answer. Further studies should be concerned not only with the prevailing relation between technology and education, but also with the ways in which various types of education and training are utilized on different kinds of jobs. We need to learn more about productivity or performance differences between people working with similar kinds of equipment, but having different amounts of education. We also need to know more about hiring and promotional practices relative to education under varying technical conditions. Studies dealing with these problems should be conducted in a wide range of technological settings, using interviews with workers, supervisors, and personnel managers as well as company records, and the results of aptitude tests. It is regrettable that the new Economics of Education has been so largely preoccupied with cost-benefit analysis. It is also regrettable that it has relied so largely on the assumption that earnings differentials between education groups reflect productivity differentials. By this assumption it has been able to bypass questions about the productivity, utilization, and adequacy of education in various work settings.

Appendix I

INTERVIEWING AND QUESTIONNAIRE DESIGN

The study was aimed at interviewing a representative sample of labor force participants residing in private households in the continental United States. In the first contact with the selected households, one family member was asked to provide information on the employment situation of all family members; this information would identify the household members to be contacted (i.e. the possible labor force participants). At the time of the actual interview, each respondent was asked a further series of questions about his work status (see page 1 of the questionnaire) in order to verify his (or her) labor force status and screen out those who actually were not eligible members of the labor force (part-time work of less than 20 hours a week was excluded). Every eligible member of the labor force was then interviewed individually.

Under no circumstances was the interviewer permitted to obtain information from another household member, instead of interviewing the employed person himself (or herself). Thus in some households no interview at all was taken, for instance households containing only retired people, housewives, students, and children. In many other households two or more interviews were conducted.

The interviews were taken in two waves, the first in May 1967 and the second in October 1967. The questions asked were identical in the two waves, and the two waves were combined in the analysis. Altogether 2,662 members of the labor force were interviewed.

The interviewing methods used were those standard with the Survey Research Center.[1] All interviews were personal interviews in the homes of respondents. The interviewers used a carefully designed and pretested questionnaire and asked questions exactly as they appeared in this questionnaire. Respondents answered in their own words, often at some length; the interviewers recorded answers as nearly verbatim as possible. The answers were then coded at the Survey Research Center in Ann Arbor.

To treat the different types of employment patterns encountered in this survey, it was necessary to use four separate interview schedules. The "A" schedule is the basic one; respondents who held the same job for the past 5 years went through only the "A" schedule. Where it was determined that a person was self-employed, the interviewer immediately administered the "S" schedule, then supplemented this with the "A" schedule for basic socio-economic information. For respondents who had a job change or a job transfer, questions in the "J" and "T" schedules respectively were asked as well as some sections of the "A" schedule.

Schedules "A" and "J," reproduced below, cover nearly all of the important questions asked; "T" is virtually identical to "J," and most questions asked in "S" are similar also.

It may be noted that Part II of "J" is practically identical to Section V of "A," but references to the effect of the job change are substituted for references to the equipment change. The answers were treated in a parellel fashion in Chapter 6.

Those currently unemployed were asked an abbreviated series of questions from "A" and "J" about their most recent job, as well as the reason for their unemployment.

In connection with the pretesting of the questionnaire, a small validation experiment was conducted in which interviews with employees in selected firms were followed by interviews with their supervisors. The object of this experiment was to assess the accuracy of the non-attitudinal information given by workers about their jobs and equipment, changes in equipment during the past 5 years and the impact of these changes on work content, employment, and pay. In all, 30 matched pairs of interviews were obtained in three firms in the face of a large number of procedural obstacles. The validation experiment was encouraging in that in a substantial majority of cases the two interviews were consistent. Where there were discrepancies, it was clear in some instances that the supervisor was not close enough to the various tasks performed by the worker to give reliable information. More often the worker seemed to be

---

[1]See Robert Kahn and Charles Cannell, *The Dynamics of Interviewing,* John Wiley, 1957; also Interviewers' Manual, Survey Research Center, Institute for Social Research, University of Michigan, Ann Arbor, 1966. (Revised Edition 1969).

the less articulate and the less accurate of the two respondents. However, it was often unclear which of the two respondents was right and which was wrong.

Consistency increased with the education of the respondent. For the most part, equipment changes, their impact on the number of people required to do the work, the abolition of jobs, changes in skill, speed, and physical effort required as well as transfers—all were reported consistently by the employee and his supervisor. The most difficult questions evidently were those pertaining to the characteristics of the worker's equipment and the nature of technical changes in the equipment, particularly equipment which the respondent did not operate himself. Most people were, however, able to provide the correct name of the equipment they were using. Although the validation experiment could not yield anything like a precise estimate of respondent accuracy, it was useful in pointing to weaknesses in the questionnaire and to problem areas which should receive special attention in interviewer training. The final questionnaire was a revised version of the one used in the validation experiment.

SℛC SURVEY RESEARCH CENTER / INSTITUTE FOR SOCIAL RESEARCH / THE UNIVERSITY OF MICHIGAN	FALL 1967 OMNIBUS / PROJECT 770 / November 1967 / DL-MT-252A / BB #44-56603	Interview No. _____ / Place codes _____ / (Do not write in above spaces)

A

(When checked, indicates that some equipment is important to R's present job)

```
┌─────────────────────────────┐
│                             │
│   1. Interviewer's Label    │
│                             │
└─────────────────────────────┘
```

2. PSU Name _____

3. Your Interview No. _____

4. Date _____

5. Town or city _____   6. Time started _____

7. (ENTER FROM COVER SHEET LISTING BOX) Sex of R _____ Age of R _____

(JOB SCREENING QUESTIONS—to be asked of this R personally—Be sure to check one of the boxes in Q.8)

8. Are you working at a full-time job now?
   (IF NO) 8a. Do you have a part-time job? (How many hours per week do you work on your (main) job?

CHECK ONE:

1. NOT WORKING AT ALL NOW	2. NOW WORKING PART-TIME LESS THAN 20 HOURS	3. PART-TIME 20 OR MORE HOURS (ENTER NO. OF HOURS _____)	4. FULL-TIME JOB

→ (BEGIN INTERVIEW)

(Include here if temporarily absent due to strike, sickness, weather, vacation, or personal reasons)

9. Have you held a job of 20 or more hours per week at any time during the past 5 years?

1. YES	5. NO	→ (TERMINATE INTERVIEW)

(IF NOW WORKING PART-TIME LESS THAN 20 HOURS)

9a. Would you like to be working 20 or more hours per week if such a job could be found?

WOULD LIKE TO BE WORKING 20 OR MORE HOURS PER WEEK	VOLUNTARILY WORKING LESS THAN 20 HOURS PER WEEK

(ASK Q's 10-11 USING THE PHRASE "a job working 20 or more hours per week")

(TERMINATE INTERVIEW)

(IF NOT WORKING AT ALL NOW)

9b. Are you unemployed, or laid-off, or retired, or what?

UNEMPLOYED LAID-OFF (ASK Q's 10-11 USING THE WORD "work")

| ☐ RETIRED |
| ☐ DISABLED PERMANENTLY |
| ☐ DISABLED, NOT PERMANENTLY |
| ☐ HOUSEWIFE |
| ☐ STUDENT |
| ☐ VOLUNTARILY IDLE |
| ☐ OTHER (EXPLAIN IN T'NAIL) |

(TERMINATE INTERVIEW)

10. During the past 4 weeks, have you been looking for (work) (a job working 20 or more hours per week)?

1. YES	5. NO	→ (SKIP TO Q. 11)

10a. What have you been doing in the last 4 weeks to find (work) (a job working 20 or more hours per week)?
_____

11. Is there any reason why you could not have taken (work) (a job working 20 or more hours per week) if it could have been found during the last week?

1. NO REASON	2. ONLY REASON WAS TEMPORARY ILLNESS	3. OTHER TEMPORARY REASON (EXPLAIN IN THUMBNAIL)	4. NONTEMPORARY REASON (EXPLAIN IN THUMBNAIL)

(BEGIN INTERVIEW)   (TERMINATE INTERVIEW)

ⓒ 1967 The University of Michigan

2

JOB SECTIONS: INTRODUCTION

(IF R IS UNEMPLOYED NOW [OR IF HE IS WORKING LESS THAN 20 HOURS PER WEEK] Q's B1-B6
REFER TO HIS MOST RECENT JOB OF 20 HOURS OR MORE -- SEE THE SPECIAL SECTION TOWARD
THE END OF THE INSTRUCTION BOOK FOR THE PATH TO FOLLOW THROUGH THE QUESTIONNAIRE FOR
THESE PEOPLE)

B1.   What is your main occupation?   (I mean your main job.) _____

_____

B2.   What kind of business is that in? _____   _____

_____

B3.   Do you work for someone else, or for yourself, or what?

| 1. SOMEONE ELSE | 3. BOTH | 2. SELF-EMPLOYED ONLY (R does not have a job of 20 or more hours working for someone else) → | (GO TO GREEN SCHEDULE S - SELF-EMPLOYED) |

B4.   Does the company you work for have plants or offices at more than one location?

| 1. YES | 5. NO |

B4a.   About how many people are employed altogether by your company or
employer?

| 0. 1-2 | 1. 3-9 | 2. 10-49 | 3. 50-99 | 4. 100-499 |

| 5. 500-999 | 6. 1000-1999 | 7. 2000-4999 | 8. 5000 AND OVER |

B4b.   About how many people are employed by your company or employer <u>at the
location</u> where you work, I mean all types of workers in all areas and
departments?

| 0. 1-2 | 1. 3-9 | 2. 10-49 | 3. 50-99 | 4. 100-499 |

| 5. 500-999 | 6. 1000-1999 | 7. 2000-4999 | 8. 5000 AND OVER |

| &. R works out of more than one location | (EXPLAIN): _____

_____

B5.   Is this more or less than it was about 5 years ago or is it about the same?

| 1. MORE | 3. SAME | 5. LESS | 8. DON'T KNOW |

B6.   Could you please tell me a little more about what <u>you</u> do on your job?

_____

_____

3

SECTION I: EQUIPMENT R OPERATES OR HELPS TO RUN ON MAIN JOB

B7.   CHECK ONE:

R IS WHITE COLLAR WORKER (SALES, OFFICE WORK OR PROFESSIONAL, BUSINESS MAN)	ANY OCCUPATION OTHER THAN WHITE COLLAR

B7a. Nowadays even people in sales and office jobs and people in the professions often use machines or help operate mechanical equipment. For example, we have in mind any equipment that has a motor, or any equipment that is run by electricity or air pressure, or anything like that. On your present job, do you ever <u>operate</u> or <u>help to run</u> any kind of mechanical equipment? (EXCLUDE ORDINARY TELEPHONES)

B7b. On your present job, do you ever <u>operate</u> or <u>help to run</u> any kind of a machine, mechanical equipment, or power tools on your job? I mean tools or machinery that have a motor or a gasoline engine, or equipment that is run by electricity or air pressure, or anything like that? (EXCLUDE ORDINARY TELEPHONES)

YES   NO ⟶ (SKIP TO Q. B20, P. 8)   YES   NO ⟶ (SKIP TO Q. B20, PAGE 8)

B8.   What kinds of equipment or machinery are these? Any other kind?

(ASK B8a - B8d ABOUT EACH PIECE OF EQUIPMENT

B8a. What is it called? (TRY TO GET EXACT NAME)	B8b. What does this equipment do?
1.	
2.	
3.	

IMPORTANT INSTRUCTIONS (to be read before beginning interviewing):

A) If R should mention the following kinds of equipment, it should not be included here in Section I, but rather should be entered in Q C2 in Section II, Page 9:

Equipment that R inspects, cleans, designs, makes, sells, installs, or repairs, etc., and equipment operated by people R supervises.

B) If R operates or helps to run many different pieces of equipment, enter several of the most important in B8a, and then try to get a general description of the rest of the equipment to enter in B8a. (And ask B8b-e with reference to this group of equipment collectively.) DO NOT spend a lot of time getting a lot of details about many individual pieces of equipment.

C) Definition of "not commonplace" as used in Column B8e: A piece of equipment is "not commonplace" (and should therefore be checked in Column B8e) if an average person would not be familiar with the details of how the equipment operates, so that it would be useful to ask the probes B14-B16 about how the equipment operates. Examples of some equipment that would be commonplace: Ordinary trucks, buses, typewriters, cash registers, adding machines, home appliances, musical instruments, etc.

B8c. What kinds of things do you do while operating or helping to run this equipment?	B8d. Do you use this (REPEAT NAME) almost constantly (AC), some of the time (ST), or very little (VL)?	B8e ENTER A ✓ FOR ANY EQUIPMENT (EXCEPT COMPUTER) THAT IS "NOT COMMONPLACE"

5 INTERVIEWER: FOLLOW DIRECTIONS IN EVERY ONE OF I, II, III, AND IV WHICH APPLIES TO R

I.   (IF JUST ONE PIECE OF EQUIPMENT IS LISTED IN B8a)

B9.   CHECK ONE   (SEE Q. B8d):

THIS EQUIPMENT IS USED:	ALMOST CONSTANTLY	SOME OF THE TIME	VERY LITTLE
	(CHECK THE BIG "A" IN THE UPPER LEFT CORNER OF THE FACE SHEET AND SKIP TO III ON NEXT PAGE)	(CHECK THE BIG "A" IN THE UPPER LEFT CORNER OF THE FACE SHEET AND ASK Q. B9a)	(ASK Q. B9b)

B9a. What proportion of your time on the job is spent working with this (REPEAT NAME OF EQUIPMENT), one-quarter, one-half, three-quarters, or what?

_____

(SKIP TO III ON NEXT PAGE)

B9b. When, or under what circumstances, do you work with the (REPEAT NAME OF EQUIPMENT)?

_____

(SKIP TO III ON NEXT PAGE)

II.   (IF MORE THAN ONE PIECE OF EQUIPMENT IS LISTED IN B8a)

B10.   CHECK ONE:
(SEE Q. B8d)

SOME OF THE EQUIPMENT IS USED ALMOST CONSTANTLY	NONE OF THE EQUIPMENT IS USED ALMOST CONSTANTLY
(CHECK THE BIG "A" IN THE UPPER LEFT CORNER OF THE FACE SHEET AND SKIP TO III ON NEXT PAGE)	(CONTINUE WITH Q. B10a)

B10a. Thinking now of all the equipment that you operate or help to run on your job, do you use at least one piece or another of this equipment:

Almost constantly	Some of the time	Very little
(CHECK THE BIG "A" IN THE UPPER LEFT CORNER OF THE FACE SHEET AND SKIP TO III ON NEXT PAGE)	(CHECK THE BIG "A" IN THE UPPER LEFT CORNER OF THE FACE SHEET AND ASK Q. B10b)	(ASK Q. B10c)

B10b. What proportion of your time on the job is spent working with this equipment, one-quarter, one-half, three-quarters, or what?

_____

(GO ON TO III ON NEXT PAGE)

B10c. When, or under what circumstances, do you work with this equipment?

_____

(GO ON TO III ON NEXT PAGE)

6

III.   (IF A COMPUTER IS LISTED IN B8a)

> B11.   Do you operate a computer, or write programs for it, or feed information to it, or what? (Tell me about it.)
>
> _____
>
> _____
>
> B12.   How long have you been working with a computer?
>
> _____

IV.   (ALL RESPONDENTS)

> B13.   CHECK ONE:

COLUMN B8e CONTAINS AT LEAST ONE CHECKMARK (i.e. some of R's equipment [other than a computer] is "not so commonplace")	COLUMN B8e DOES NOT CONTAIN ANY CHECKMARKS (i.e. all of R's equipment [other than a computer] is "commonplace")
(ASK Q. B14 - B16 WITH REFERENCE TO THE CHECKED EQUIPMENT)	(SKIP TO Q. B 17 ON PAGE 7)

B14.   Is this equipment the kind where you have to control absolutely everything it does yourself, or does it do some things automatically?

_____

_____

1. SOME AUTOMATIC FEATURES	3. NONE HAS AUTOMATIC FEATURES (SKIP TO Q. B17, PAGE 7)	DON'T UNDERSTAND Q. OR DON'T KNOW WHETHER HAS ANY AUTOMATIC FEATURES

> B14a.   For example, does it automatically feed a new piece after another is finished, or perform two or more operations one after another without your doing something, or automatically control the flow of work, or signal when a process is complete, or anything like that?
>
> _____

2. SOME AUTOMATIC FEATURES	4. NONE HAS AUTOMATIC FEATURES (SKIP TO Q. B17, PAGE 7)	8. DON'T KNOW (SKIP TO Q. B17, PAGE 7)

B14b.   What kinds of things does it do automatically? _____

_____

_____

7

B15.   Is any of this equipment you work with connected up to any kind of tape or numeric control unit, or to a computer? (Tell me about it.)

_____

_____

_____

| 1. SOME IS CONNECTED TO CONTROL OR COMPUTER | 5. NONE IS CONNECTED TO CONTROL OR COMPUTER (SKIP TO Q. B16) |

B15a.   Do you happen to know whether the computer (control unit) keeps track of time, or counts things, or records things like speed, size, shape, temperature, weight, or anything like that? ...What does it do? ...Anything else it does?

_____

_____

_____

B15b.   About how long have you been working with pieces of equipment that are connected to a computer or numeric control unit?

_____

_____

B16.   Is any of the equipment you work with connected up to another piece of equipment by conveyors or pipes, or electrical connections, or something like that - so that they work together as a unit? (Tell me about it.)

_____

_____

B17.   Suppose someone else were to do the same job as yours, how much education and special training do you think he would have to have to do what you do on your job?

_____

_____

B18.   Thinking now about just the equipment or machinery you use on your job, how long would it take someone with the same general education and training as you to learn how to work properly with this equipment?

_____

_____

8

B19. Some people tell us that the equipment they work with is like a friend that helps them to do something. Other people say that their equipment is more like a foe that they have to struggle with, or that is even running their lives. How do you feel about the equipment on your job?

_____

_____

_____

B20. Do you have much chance to talk to other people during the time you are working, or does your location or the kind of work you do keep you from talking with others while you work?

_____

_____

B21. In some jobs, how your work is organized depends mostly on you, yourself. In others, it's pretty much determined by the equipment you work with, a production line, or by a number of people who have to work together. How is it with your job?

_____

_____

_____

9   SECTION II: EQUIPMENT THAT R DOES <u>NOT</u> OPERATE OR HELP TO RUN

INCLUDE HERE EQUIPMENT THAT R INSPECTS, CLEANS, DESIGNS, MAKES, SELLS, INSTALLS, REPAIRS, ETC., AND EQUIPMENT OPERATED BY PEOPLE R SUPERVISES.

C1. Sometimes a change in some equipment can be important for a person's work even if he doesn't operate it, or operates it very little. This might be equipment in your own department, in other departments, or even in other companies. It might be equipment that you inspect, clean, design, make, sell, install, or repair. Is there any such equipment which is important to you because a <u>change</u> in that equipment would affect how you do your work?

☐ YES, SOME EQUIPMENT IS IMPORTANT   ☐ NO
   (CHECK THE BIG "A" IN THE UPPER LEFT   (SKIP TO SECTION III,
   CORNER OF THE FACE SHEET AND CONTINUE   PAGE 13)
   WITH Q.C2)

C2. What kinds of equipment or machinery are these? Any other kind?

(ASK Q. C2a-d ABOUT EACH PIECE OF EQUIPMENT)

C2a. What is it called? (TRY TO GET EXACT NAME)	C2b. What does it do?

(IF R MENTIONS A COMPUTER)

10

C3. Could you tell me a little bit more about the kinds of things this computer is used for?

_____
_____
_____

C4. How long has a computer been important in relation to your work?

_____
_____
_____

(CONTINUE WITH SECTION III, NEXT PAGE)

C2c. In what way is this equipment important to your work?	C2d. Where is this equipment located - in your own department, in another department of your company, in another company, or where?	
_____	_____	EQUIPMENT R INSPECTS, CLEANS, DESIGNS, MAKES, SELLS, INSTALLS, REPAIRS, ETC., IS BY DEFINITION IN HIS DEPARTMENT.
_____	_____	
_____	_____	
_____	_____	
_____	_____	
_____	_____	
_____	_____	
_____	_____	
_____	_____	

(IF R MENTIONS A COMPUTER, GO TO Q. C3, TOP OF THIS PAGE)

(IF R DOES NOT MENTION A COMPUTER, GO TO SECTION III, PAGE 13)

11

C3. Is this equipment the kind where the operator has to control absolutely everything it does himself, or does it do some things automatically?

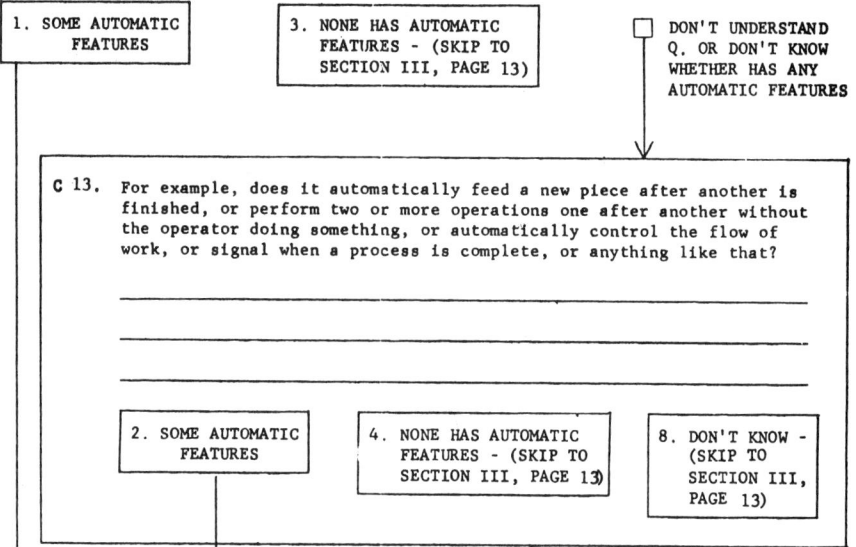

_____

_____

_____

| 1. SOME AUTOMATIC FEATURES | 3. NONE HAS AUTOMATIC FEATURES - (SKIP TO SECTION III, PAGE 13) | ☐ DON'T UNDERSTAND Q. OR DON'T KNOW WHETHER HAS ANY AUTOMATIC FEATURES |

C 13. For example, does it automatically feed a new piece after another is finished, or perform two or more operations one after another without the operator doing something, or automatically control the flow of work, or signal when a process is complete, or anything like that?

_____

_____

_____

| 2. SOME AUTOMATIC FEATURES | 4. NONE HAS AUTOMATIC FEATURES - (SKIP TO SECTION III, PAGE 13) | 8. DON'T KNOW - (SKIP TO SECTION III, PAGE 13) |

C 3b. What kinds of things does it do automatically? _____

_____

_____

_____

12

C 4.   This equipment that's important for your job, even though you don't operate it;
is any of it connected up to any kind of tape or numeric control unit, or to a
computer? (Tell me about it.)

_____

_____

_____

```
┌─────────────────────────┐        ┌──────────────────────────┐
│ 1. SOME IS CONNECTED TO  │        │ 5. NONE IS CONNECTED TO  │
│    CONTROL OR COMPUTER   │        │    CONTROL OR COMPUTER   │
└─────────────────────────┘        │    (SKIP TO Q.C13 BELOW) │
            │                      └──────────────────────────┘
            ▼
```

C4a.   Do you happen to know whether the control unit or computer keeps track of
time, or counts things, or records things like speed, size, shape, temperature,
weight, or anything like that? ...What does it do? ...Anything else it does?

_____

_____

_____

C4b.   About how long has equipment connected up to a computer or numeric control unit
been important for your job?

_____

_____

C5.   Are any of the pieces of automatic equipment which are important for your job
connected up to another piece of equipment by conveyors, or pipes, or electrical
connections, or something like that - so that they work together as a unit? (Tell
me about it.)

_____

_____

_____

_____

13

SECTION III: EQUIPMENT CHANGE

D1. How long have you worked for your present employer? _____ (years)

   D1a. Counting only your main job, how many different companies or employers have
   you worked for in the last 5 years, including your present employer?

   _____ (ENTER NUMBER)

| 1. ONE EMPLOYER | 2. MORE THAN ONE (GO TO YELLOW SCHEDULE J - JOB CHANGE) | Disregard here any previous self-employment |

D2. Have you been transferred or reassigned to another section or department or to a
different plant of your company at any time during the past 5 years (not counting
very temporary shifts)?

| 3. YES, TRANSFER OR REASSIGNMENT (GO TO BEIGE SCHEDULE T - TRANSFER) | 4. NO TRANSFER OR REASSIGNMENT | GO TO D3 OR D4 |

(IF R DOES NOT NOW OPERATE OR HELP TO RUN ANY EQUIPMENT)

D3. Was there a time during the last 5 years when you operated or helped to run
any equipment on your job?

_____

   [YES] ⟶ SKIP TO Q. D6 BELOW   [NO] ⟶ SKIP TO Q. D11, PAGE 15

D4. During the last 5 years was there any change at all in any equipment or power
tools which you yourself have operated or helped to run on your job?

_____

_____

D5. Was there any change in equipment which is directly connected to any equipment
which you have operated or helped to run?

_____

☐ YES, SOME CHANGE MENTIONED
   IN EITHER D4 OR D5.

☐ NO CHANGES MENTIONED
   IN EITHER D4 OR D5.
   (SKIP TO Q. D11, PAGE 15)

D6. Thinking of the most important change during the past 5 years in the machinery
or equipment which you have operated or helped to run, I mean the change which
had the greatest effect on you and your job, when did that occur?

_____

14

D7.   What kind of change in equipment was this?   (What equipment was changed?   In what way was it changed?)

_____

_____

_____

D8.   Would you say that what <u>you</u> had to do on your job was changed quite a bit by the new equipment, was it changed somewhat, or did you keep on doing nearly the same thing in nearly the same way?

1. CHANGED QUITE A BIT	3. CHANGED SOMEWHAT	5. KEPT ON DOING NEARLY THE SAME THING

(CHECK BOX 1 AT THE TOP OF PAGE 17 AND CONTINUE WITH Q.D8a)   (SKIP TO Q.D9)

D8a.   In what ways was your work changed? _____

_____

_____

D9.   Would you say that in general the new equipment required the people working with it to do <u>more</u> or <u>less</u> hand work or manual tasks than the old, or was there no change in this respect?

_____

_____

(IF CHANGE)

D9a.   What kind of hand work or manual tasks was there (more/less) of?

_____

D9b.   Was <u>your</u> job affected in this respect? (More or less hand work or manual tasks?)

_____

_____

_____

(IF OWN JOB   D9c.   Did this mean a large change or a small change in your AFFECTED)   own work? _____

D10.   Were more people or fewer people needed to put out the <u>same amount of work</u> per day, or was there no change in this respect? _____

_____

(IF MORE OR FEWER)
D10a.   Was this a large change or a small change? _____

15

D11. Now, thinking again about equipment that you don't operate but which is important for your work - I mean equipment in your company or even in other companies - during the last 5 years was there any change at all in equipment having an effect on how you do your work?

| 1. YES, SOME CHANGE | | 5. NO | → | (CHECK BOX 4 AT TOP OF PAGE 17 [UNLESS BOX 1 IS ALREADY CHECKED] AND SKIP TO Q. D18, PAGE 16) |

D12. Thinking of the most important change during the past 5 years in this equipment that affects how you do your work, when did that occur?

_____

D13. What kind of change in equipment was this? (What equipment was changed? In what way was it changed?)

_____

_____

_____

D14. Would you say that what you had to do on your job was changed quite a bit by the difference in equipment, was it changed somewhat, or did you keep on doing nearly the same thing in nearly the same way?

| 1. CHANGED QUITE A BIT | 3. CHANGED SOMEWHAT | 5. KEPT ON DOING NEARLY THE SAME THING |

D14a. In what ways was your work changed?

(CHECK BOX 4 AT TOP OF PAGE 17 [UNLESS BOX 1 IS ALREADY CHECKED] AND CONTINUE WITH Q. D15)

_____

_____

_____

_____

_____

_____

D15. Was this equipment that was changed located in your company or in another company?

_____

| IN R's COMPANY (CONTINUE WITH D16, PAGE 16) | OUTSIDE R's COMPANY (SKIP TO D18, PAGE 16) |

D14b. Was this equipment that was changed located in your company or in another company?

_____

| IN R'S COMPANY (CHECK BOX 2 [UNLESS BOX 1 IS ALREADY CHECKED] AND CONTINUE WITH D16) | OUTSIDE R's COMPANY (CHECK BOX 3 AT THE TOP OF PAGE 17 [UNLESS BOX 1 IS ALREADY CHECKED] AND SKIP TO Q. D18, PAGE 16) |

16

D16. Would you say that in general the new equipment required the people working with it to do <u>more</u> or <u>less</u> hand work or manual tasks than the old, or was there no change in this respect?

_____

(IF CHANGE)

D16a. What kind of hand work or manual tasks was there (more/less) of? _____

_____

D16b. Was <u>your</u> job affected in this respect? (More or less hand work or manual tasks?)

_____

_____

_____

    (IF OWN JOB   D16c. Did this mean a large change or a small change in your
    AFFECTED)   own work?

                                                              _____

D17. Were more people or fewer people needed to put out the <u>same amount of work</u> per day, or was there no change in this respect?

_____

(IF MORE
OR FEWER)   D17a. Was this a large change or a small change? _____

D18. People's jobs often change even if there are no changes in the equipment they use. Have there been any changes in your work during the past five years that we have not talked about?

_____

(IF YES)   D18a. What kinds of changes in your work were these?

_____

_____

_____

          D18b. Thinking about these things you just told me about, was what you had to do on your job changed quite a bit, changed somewhat, or did you keep on doing nearly the same thing in nearly the same way?

| 1. CHANGED QUITE ·A BIT | 3. CHANGED SOMEWHAT | 5. KEPT ON DOING NEARLY SAME THING |

NOTE: By the time you get here, you should have checked <u>one</u> of the boxes at the top of page 17. If necessary, refer to Q's D8, D1̄1, D14, D14b.

17

D19.   CHECK <u>ONE</u> BOX: <u>IF BOX 1 IS ALREADY CHECKED, DO NOT</u> CHECK BOX 2, 3 OR 4.

R's job changed "quite a bit" or "somewhat"
by change in equipment which is:

Operated by R	<u>Not</u> operated by R, but located <u>in</u> R's company	N_t operated by R, and located <u>outside</u> R's company	No change in equipment, or change affects R's job very little

CHECK <u>ONE</u> BOX ONLY:

Box 1	Box 2	Box 3	Box 4
(CONTINUE WITH SECTION IV AND ASK ABOUT THE CHANGE IN <u>EQUIP-MENT WHICH R OPERATES</u>)	(CONTINUE WITH SECTION IV)	(SKIP TO SECTION V, PAGE 20)	(SKIP TO SECTION VI, PAGE 23)

SECTION IV: WHAT HAPPENED WHEN EQUIPMENT WAS CHANGED IN R's COMPANY?

Now we are interested in how the company handled the change-over to the new equipment (that you work with/that affects your job) and how you felt about the change.

E1.   How much notice did your employer give you about the change he was making in machinery or equipment?

_____

E2.   Do you feel you had enough advance notice, or was it insufficient?

_____

(IF INSUFFICIENT)

E3.   Why is that?   _____

_____

_____

E4.   Was the change-over planned by the company alone, or did employees participate in the planning, or did a union have some say also, or what?

_____

_____

18

E5. Did the change-over to the new equipment affect the number of people needed to do the work in your section? (Were more or fewer people needed after the change-over?)

_____

| 1. MORE NEEDED | 5. FEWER NEEDED | 0. NOT AFFECTED | → (SKIP TO Q.E6) |

E5a. Why is that? _____

_____

E5b. Were a lot (more/fewer) people needed to do the work, in your section, or just a small number?

_____

E6. Did any of the workers in your section become unemployed or work shorter hours for any period at all as a result of the change-over to the new equipment? (Which?)

_____

(IF YES) E6a. Were quite a few or just a small number of people involved?

_____

E7. Was your own job abolished by the new equipment?

_____

E8. Were you laid-off, or did you stop working, or work shorter hours at that time? (What did you do?)

_____

(IF LAID-OFF OR   E8c. About how long were you out of work at that particular
STOPPED WORKING)   time?

_____

E9. Were there any changes in your seniority rights or fringe benefits as a result of the change in equipment? (Which?)

_____

E10. Was your pay raised by the introduction of the new equipment, or lowered, or was it unaffected?

| 1. RAISED | 5. LOWERED | 3. UNAFFECTED |

19

E11. In order to work with the new equipment, did you have to learn anything new or did you acquire any new skills?

| 1. YES | | 5. NO |——→(SKIP TO Q. E18 BELOW)

E11a. How did you acquire the new skill or knowledge - did you learn it by yourself on the job? Did someone train you on the job? Or did you take a formal training program or course?

> CHECK AS MANY BOXES AS APPLICABLE, THEN ASK CONTINGENT QUESTIONS FOR EACH TRAINING PROGRAM

☐ FORMAL TRAINING PROGRAM OR COURSE -- ASK Q's E12 - E17.

☐ TRAINED ON JOB -- ASK Q's E16 - E17.

☐ LEARNED BY SELF -- ASK Q.E17.

E12. Where did you take this training program or course - did the company give it, or a union, did you go to a school or university, take a correspondence course, or what?

_____

_____

E13. Was this voluntary on your part or were you asked to do it?

_____

_____

E14. Who paid for it, or was it free?   PAID FOR OR PROVIDED BY:

☐ FREE | 1. RESPONDENT | 2. EMPLOYER | 3. GOVERNMENT |

E14a. Who provided it? | 4. EQUIPMENT PRODUCER | OTHER: _____

E15. Did you take it before the change-over to the new equipment, or after, or just about at the same time? | 1. BEFORE | | 5. AFTER | | 3. ABOUT THE SAME TIME |

E16. How long a program or course was it - how many hours or days or weeks?

_____ HOURS ALTOGETHER

> IF ANSWER IN TERMS OF DAYS OR WEEKS, PROBE TO FIND TOTAL NUMBER OF HOURS

E17. What kinds of things did you learn? _____

_____

(IF DID NOT TAKE FORMAL TRAINING PROGRAM OR COURSE -- SEE Q. E11a ABOVE)

E18. At the time of the change-over to the new equipment, did you have the _opportunity_ to take a formal training program of some kind offered by your employer, by a union, by the government or by anyone else? _____

_____

20

E19. Altogether what helped you the most in getting used to the change in your work?
(Anything else?)

_____

_____

_____

E20. Was there anything which made it hard for you to get used to the change? (What was it?)

_____

_____

_____

SECTION V:　CHANGE IN JOB CONTENT FROM EQUIPMENT CHANGES IN AND OUTSIDE OF R'S COMPANY

Now we want to discuss with you some of the details of how your job was changed by the change in machinery or equipment.

F1. Is your physical work increased by the new equipment, or reduced, or does the new equipment make no difference?

| 1. INCREASED | 5. REDUCED | 3. NO DIFFERENCE |

F2. Is the speed with which you have to work increased by the new equipment, or slowed down, or does the new equipment make no difference?

| 1. INCREASED | 5. SLOWED DOWN | 3. NO DIFFERENCE |

F3. Is more skill required of you with the new equipment or less skill?

| 1. MORE SKILL | 5. LESS SKILL | 3. NO DIFFERENCE |

F4. Did the physical surroundings at work - things like light, heat, noise, space, ventilation - become more pleasant on account of the new equipment, or less pleasant?

| 1. MORE PLEASANT | 5. LESS PLEASANT | 3. NO DIFFERENCE |

F5. Since the new equipment was introduced, do you have occasion to talk more often with other people at work, or are you more alone?

| 1. MORE OFTEN | 5. MORE ALONE | 3. NO DIFFERENCE |
(SKIP TO Q. F6, PAGE 21)

F5a. Do you find this more pleasant, or less pleasant, or does it make no difference?

| 1. MORE PLEASANT | 5. LESS PLEASANT | 3. NO DIFFERENCE |

21

F6.   Do you have more reports to write or more records to keep since the new equipment came in, or less, or is there no difference?

| 1. MORE | 5. LESS | 3. NO DIFFERENCE | 0. NOT APPLICABLE |

F6a.   Why is that? _____

_____

_____

F7.   Is there more danger of personal injury with the new equipment or less danger?

| 1. MORE DANGER | 5. LESS DANGER | 3. NO CHANGE |

F8.   With the new equipment, is the opportunity to learn new things about the work greater than before or less?

| 1. MORE OPPORTUNITY | 5. LESS OPPORTUNITY | 3. NO CHANGE |

F9.   Is there more need for planning, judgment or initiative on _your_ part after the introduction of the new equipment, or less need?

| 1. MORE NEED | 5. LESS NEED | 3. NO DIFFERENCE |

F9a.   In what ways has your job changed in these respects (planning, judgement, initiative)? (Anything else?)

_____

_____

F10.   In some jobs, how your work is organized depends mostly on you, yourself. In others it's pretty much determined by the equipment you work with, by a production line, or by a number of people who have to work together. Since the new equipment came in do you feel that _you_ have more influence on how your work goes, or less influence?

| 1. MORE INFLUENCE | 5. LESS INFLUENCE | 3. NO CHANGE (SKIP TO Q.F11) |

F10a. Why is it that you have (more/less) influence on how your work goes?

_____

_____

_____

22

F11. If you made a mistake on your work, would it likely be more serious <u>with</u> the new equipment, or less serious?

| 1. MORE SERIOUS | 5. LESS SERIOUS | 3. NO CHANGE |

F11a. Why is that? _____

_____

(ASK ONLY IF R IS ENGAGED IN PRODUCTION WORK)

F12. Would you say that with the new equipment the quality of the product you make is more dependent on how well you do your work, or less dependent?

| 1. MORE DEPENDENT | 5. LESS DEPENDENT | 3. NO DIFFERENCE |

(ASK EVERYONE)

F13. Would you say that what you do on your job is more closely supervised after the new equipment came in, or less closely supervised?

| 1. MORE CLOSELY SUPERVISED | 5. LESS CLOSELY SUPERVISED | 3. NO CHANGE |

F14. Sometimes the use of new equipment also affects the chances of unemployment or being laid off. Is the work more steady on your job after the introduction of the new equipment, or less steady?

| 1. MORE STEADY | 5. LESS STEADY | 3. NO CHANGE |

F15. For a person like yourself, are the chances of moving up to a better job greater with the new equipment or less?

| 1. GREATER CHANCE | 5. LESS CHANCE | 3. NO DIFFERENCE |
| | | (SKIP TO Q.F16) |

F15a. Why is that?_____

_____

_____

_____

23

F16. Would you say that in general your work is more interesting <u>after</u> the introduction of the new equipment, or is it more monotonous?

| 1. MORE INTERESTING; LESS MONOTONOUS | 5. LESS INTERESTING; MORE MONOTONOUS | 3. NO CHANGE (SKIP TO Q.G2 BELOW) |

F16a. In what ways is your work (more/less) interesting (monotonous)?

_____

_____

(SKIP TO Q. G2 BELOW)

SECTION VI: PAST AND EXPECTED CAREER CHANGES

G1. In some jobs, how your work is organized depends mostly on you, yourself. In others it's pretty much determined by the equipment you work with, by a production line, or by a number of people who have to work together. In comparison with 5 years ago, do you feel that <u>you</u> <u>now</u> have more influence on how your work goes, or less influence?

| 1. MORE INFLUENCE | 5. LESS INFLUENCE | 3. NO CHANGE (SKIP TO Q. G2) |

G1a. Why is it that you now have (more/less) influence on how your work goes?

_____

_____

G2. We have asked about changes in the equipment you have worked with. Looking now to the future, do you expect that during the next few years the company you work for will modernize its machinery and equipment?

| 1. WILL | 2. PROBABLY WILL | 3. MIGHT; UNCERTAIN | 5. WILL NOT (SKIP TO Q.G3) |

G2a. What kinds of changes might these be? _____

_____

_____

G2b. How might they affect your work? _____

_____

_____

24

G3. In your opinion, what should unions do, if anything, when a company decides to introduce new machinery?

_____   _____

_____

G4. In general, would you say that automation is a good thing for people doing your kind of work, or does it cause problems, or doesn't it make any difference?

_____

_____

G4a. Why do you say so? _____

_____

_____

G5. On the whole, do you feel that the work on your present job is drudgery, or is it all right, or do you enjoy your work?

_____

G6. What are the things you like best about your job (if any)? _____

_____

G7. What are the things you like least about your job (if any)? _____

_____

G8. Taking everything into consideration, how satisfied are you with your job now as compared to five years ago - are you more satisfied, about the same, or less?

| 1. MORE SATISFIED | | 3. ABOUT THE SAME | | 5. LESS SATISFIED |

G9. Do you think there is any chance that you will change employers in the next few years?   (IF ANY CHANCE OF A CHANGE) Would you say you definitely will change, you probably will, or are you uncertain?

_____

_____

| 1. DEFINITELY WILL CHANGE | 2. PROBABLY WILL | 3. POSSIBLY, UNCERTAIN | 5. NO CHANCE OF A CHANGE (SKIP TO Q.G10) |

G9a. What (if any) would be the advantages of staying with your present employer?

_____

_____

G9b. What (if any) might be the advantages of changing jobs?

_____

25

G10.   Looking back to about 5 years ago -- did you do the same general kind of work then as you do now?

1. YES	5. NO

(SKIP TO G.11)

> G10a.   Could you tell me a little bit about what kind of work you did 5 years ago?
> _____
> _____
> _____

G11.   Were you promoted at any time during the past 5 years?

1. YES	5. NO ──▶(SKIP TO G.12)

> G11a.   How long ago was that?
> _____ (YEARS AGO)

G12.   Looking back 5 years, has your annual income from your job increased, stayed about the same, or has it gone down?

1. INCREASED	3. STAYED THE SAME	5. GONE DOWN

(SKIP TO G.13)

> G12a.   Could you tell me by about how much your annual income from your job is (higher/lower) than it was 5 years ago? (We don't need an exact figure, just an approximate amount.)
> _____ per year

G13.   And how about the next few years - do you expect that your income from your job will rise a good deal, rise slightly, remain about the same, or go down?

1. RISE GOOD DEAL	2. RISE SLIGHTLY	3. STAY THE SAME

5. GO DOWN	8. DON'T KNOW

G13a. Why do you think so? _____
_____
_____

26

G14. Is there some sort of union or professional society or association for people in your line of work?

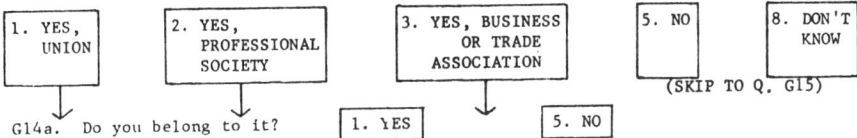

1. YES, UNION	2. YES, PROFESSIONAL SOCIETY	3. YES, BUSINESS OR TRADE ASSOCIATION	5. NO	8. DON'T KNOW

(SKIP TO Q. G15)

G14a. Do you belong to it? | 1. YES | | 5. NO |

G15. Have you lived here in ...(COUNTY)... all the time during the past 5 years or did you move here during the past 5 years?

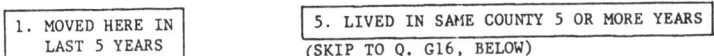

1. MOVED HERE IN LAST 5 YEARS	5. LIVED IN SAME COUNTY 5 OR MORE YEARS

(SKIP TO Q. G16, BELOW)

G15a. When was that? _____

G15b. Where did you come from, when you moved here?

_____
    (COUNTY)   (STATE)

G15c. Why did you move then? _____

_____

G15d. Was that your only move from one city or town to another, or did you make other moves during the past 5 years?

_____

G16. Do you think there is any chance that you will move away from ...(COUNTY)... in the next few years? (IF ANY CHANCE) Would you say you definitely will move, you probably will, or are you uncertain?

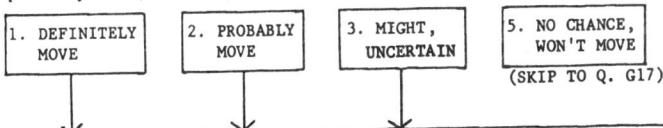

1. DEFINITELY MOVE	2. PROBABLY MOVE	3. MIGHT, UNCERTAIN	5. NO CHANCE, WON'T MOVE

(SKIP TO Q. G17)

G16a. Why are you thinking of moving? _____

_____

27

G17. Have you ever been unemployed for a week or more?

| 1. YES | | 5. NO | ⟶ (SKIP TO SECTION VII, BELOW) |

G17a. When was the last time you were unemployed for a week or more?

☐ IN LAST 5 YEARS

| NOT IN LAST 5 YEARS |

(SKIP TO SECTION VII, BELOW)

G17b. During the last 5 years, did you have several spells 'of unemployment lasting a week or more, or only one or two?

_____

G17c. Did any of your spells of unemployment last more than a month?

| 1. YES | | 5. NO | ⟶ (SKIP TO SECTION VII, BELOW) |

G17d. Why were you unemployed at that time? _____

_____

_____

_____

SECTION VII: PERSONAL DATA

H1. Are you married, single, or what?

| 2. MARRIED | 1. SINGLE | 3. WIDOWED | 4. DIVORCED | 5. SEPARATED |

H2. Do you have any children under 18 in your family living here?

☐ YES   ☐ NO ⟶ (SKIP TO Q.H3, NEXT PAGE)

H2a. How many? _____

H2b. How old are they? _____

28

H3. How old are you? _____

H4. Where did you grow up? _____
                              (STATE OR FOREIGN COUNTRY)

H5. Was that on a farm, or in a city, or a suburb of a large city, or what?
     (IF CITY) About how large a city was it?

     ☐ 1. FARM   ☐ 7. 50,000-99,999

     ☐ 2. RURAL NONFARM   ☐ 8. 100,000 and over

     ☐ 3. SUBURB OF LARGE CITY
           (with pop. 50,000 or over)   Comments: _____

     ☐ 4. CITY UNDER 2,000   _____

     ☐ 5. CITY 2,000-9,999   _____

     ☐ 6. CITY 10,000-49,999   _____

H6. How many grades of school did you finish? _____

     ☐ 12 OR MORE GRADES OF SCHOOL, OR   ☐ 1. 11 OR LESS GRADES OF SCHOOL,
       FINISHED EQUIV. OF HIGH SCHOOL   DID NOT FINISH EQUIV. OF
                                                 HIGH SCHOOL
                                                (SKIP TO Q.H7, BELOW)

     H6a. Have you been to college? ☐ YES   ☐ 2. NO ──→ (SKIP TO
                                         ↓   QH7, BELOW)
     H6b. How many years? _____

     H6c. Do you have a college degree? ☐ 4. YES   ☐ 3. NO ──→ (SKIP TO
                                              ↓   Q.H7, BELOW)
     H6d. What degree(s) do you have? _____

     H6e. In what field(s) is (are) your degree(s)? _____

(IF ATTENDED HIGH SCHOOL)
H7. Did you have any vocational courses in high school? _____

     (IF YES) H7a. What courses were they? _____
               _____
               _____

29

H8. Have you had any (other) technical or vocational training that we haven't talked about yet? We have in mind things like technical training in electronics, drafting, nursing school, barber school, things like that? (How about any apprenticeships leading to journeyman status? ...Any company training program that lasted six weeks or longer? ...Vocational training in the armed forces? Anything else?)

| 1. YES | 5. NO SUCH TRAINING ———→(SKIP TO Q. H9, BELOW) |

(LIST EACH TRAINING PROGRAM AND ASK Q's H8a - H8g FOR EACH PROGRAM MENTIONED)

H8a. For what kind of work were you trained?			
H8b. What kind of training was it, classes or home study or what?			
H8c. Was it given by a public high school, a private vocational school the company you worked for, or what?			
H8d. Who paid for it?			
H8e. How long ago did you receive it (YEARS)?			
H8f. How long did the training last altogether? (How many hours, weeks, months?) (WHICH?)			
H8g. Did you finish the training program?	1. YES 5. NO	1. YES 5. NO	1. YES 5. NO

H9. Through your previous experience and training, have you built up some skills that you would like to be using, but can't on your present job?

| 1. YES | 5. NO ———→ (SKIP TO Q.H10, PAGE 30) |

H9a. What are they? _____

_____

_____

30

H10. In connection with your future work do you feel that it would be useful for you to get additional education or some kind of training, or is there no need for it?

| 1. USEFUL, SOME NEED | | 5. NO NEED, NOT USEFUL | (SKIP TO Q.H11, BELOW) |

H10a. Why would this be useful for you? _____

_____

H10b. Do you think that you might get such education or training in the next few years, or is the idea impractical?

| 1. MIGHT GET EDUCATION OR TRAINING | 5. IMPRACTICAL TO GET |

(SKIP TO QH11, BELOW)

H10c. Why is that? _____

(INTERVIEWER: HAND INCOME FLASH CARD TO R)

H11. Can you tell me about how much money you (R) made on your own main job last year? I mean before taxes or any deductions. Just tell me the letter beside the right bracket.

| A. 0. UNDER $2000 | B. 1. $2000-2999 | C. 2. $3000-3999 | D. 3. $4000-4999 |

| E. 4. $5000-5999 | F. 5. $6000-7499 | G. 6. $7500-9999 |

| H: 7. $10,000-14,999 | I. 8. $15,000-24,999 | J. 9. $25,000 AND OVER |

H12. These are all the questions that I have. At the conclusion of this survey we can send you some of our findings, without charge, if you will send in this card. (HAND REPORT REQUEST CARD TO R.) Thank you very much for your help in this project.

H13. Time interview ended: _____

COPY YOUR INTERVIEW NUMBER ONTO OTHER SCHEDULES AS NECESSARY.

BY OBSERVATION ONLY: Sex of Respondent: ☐ MALE   ☐ FEMALE

Race: ☐ WHITE   ☐ NEGRO   ☐ OTHER (SPECIFY): _____

Did the respondent understand the questions and answer readily, or did he have some difficulty understanding and answering? (NOT COUNTING LANGUAGE DIFFICULTY)

| 1. R COULD UNDERSTAND AND ANSWER QUESTIONS SATISFACTORILY | 5. R WAS SLOW TO UNDERSTAND AND HAD DIFFICULTY ANSWERING QUESTIONS |

Thumbnail Sketch

Survey Research Center
The University of Michigan
Project 770
November 1967
DL-MT-252B
BB #44-566031
Approv. Exp. 6-68

J

JOB CHANGE

INTERVIEWER'S LABEL

Your Interview Number _____ _____

PART I: WHAT HAPPENED WHEN R CHANGED JOBS

J1. During the last five years, did you leave (any of) your former employer(s) because
he changed to new machinery or because of automation on your former job -- did
that have anything to do with your leaving?

| 1. YES | | 3. IN SOME WAYS | | 5. NO |

(SKIP TO Q.J6, PAGE 2)   (SKIP TO Q.J6, PAGE 2)

J2. What kind of work did you do on your previous job? J2a. Could you please tell
me a little bit more about what you did on your job?

_____

_____

_____

_____

J3. Why did you leave? _____

_____

_____

J4. Was the place where you worked cutting back production or reducing the work
force at the time you left your previous job?

_____

(IF YES) J4a. Why was it that the work force or production was being
reduced where you worked? _____

_____

J5. Were you unemployed for a while at that time?

| 1. YES | | 5. NO |——> (SKIP TO Q.J22, PAGE 3)

J5a. About how long were you out of work at that particular time?

_____ (WEEKS)

(SKIP TO Q.J22, PAGE 3)

2

J6. When did that happen - how many years ago? _____ (YEARS)

J7. What kind of work did you do in the company you left? J7a. Could you please tell me a little bit more about what you did on your job?

_____

_____

_____

J8. How long had you been employed by the company you left? _____ (YEARS)

J9. What kind of a change in equipment or automation was this which made you change jobs?

_____

_____

_____

J10. We are interested in how your former employer handled the change-over to the new equipment and how you felt about it How much notice did your former employer give you about the change he was making in machinery or equipment?

_____

J11. Do you feel you had enough advance notice, or was it insufficient?

_____

(IF INSUFFICIENT) J12. Why is that? _____

_____

J13. Was the change-over planned by the company alone, or did employees participate in the planning, or did a union have some say also, or what?

_____

J14. Did the change-over to the new equipment affect the number of people needed to do the work in your section? (Were more people or fewer needed after the change-over?)

1. MORE NEEDED	5. FEWER NEEDED	0. NOT AFFECTED
		(SKIP TO Q.J17, PAGE 3)

J15. Why is that? _____

_____

_____

J16. Were a lot (more/fewer) people needed to do the work in your section, or just a small number (more/fewer)?

_____

3

J17. Did any of the people you formerly worked with become unemployed or work shorter hours for any period at all as a result of the change-over to the new equipment? (Which?)

_____

(IF YES) J17a. Were quite a few, or just a small number of people involved?

_____

J18. Was <u>your own</u> job abolished by the new equipment? _____

_____

J19. Were you laid off, or did you yourself decide to quit at that time, or what?

_____

J20. Did you become unemployed, or stop working, or work shorter hours for any period of time? (Tell me about it.)

_____

(IF UNEMPLOYED OR   J21. About how long were you out of work at that particular time?
STOPPED WORKING)
_____ (WEEKS)

J22. What changes were there in your seniority rights or fringe benefits when you changed jobs? (Any other changes?)

_____

J23. Was your pay increased when you changed jobs, or was it lowered, or was there very little change?

| 1. INCREASED | 5. LOWERED | 3. VERY LITTLE CHANGE |

J24. When you changed jobs, did you move from one town to another, or were both jobs in the same location?

_____

(IF MOVED) J24a. About how many miles did you move? _____ (MILES)

J25. At the time you left, about how many people were employed by the company at the location where you worked, I mean all types of workers in all departments?

| 0. 1-2 | 1. 3-9 | 2. 10-49 | 3. 50-99 | 4. 100-499 |

| 5. 500-999 | 6. 1000-1999 | 7. 2000-4999 | 8. 5000 AND OVER |

J26. Was that more or less than it was a couple of years before you left, or don't you know?

| 1. MORE | 5. LESS | 3. SAME | 8. DON'T KNOW |

4

J27. Now, still thinking about your old job that we have just been talking about - was there any equipment which you <u>operated</u> or <u>helped to run</u> on your old job? (EXCLUDE <u>ORDINARY</u> TELEPHONES)

| 1. YES | | 5. NO | → (SKIP TO Q. J40, PAGE 7) |

(Equipment that R inspected, cleaned, designed, made, sold, installed, or repaired, etc., and equipment operated by people R supervised, should be entered in Q. J41. See instructions A and B at the top of Page 4, in the A Schedule.)

J28. What kinds of equipment or machinery were these? Any other kind?

(ASK J29-J32 ABOUT EACH PIECE OF EQUIPMENT)

J29. What was it called? (TRY TO GET EXACT NAME)	J30. What did this equipment do?

(IF A COMPUTER IS LISTED IN J29)

J33. Did you operate a computer, or write programs for it, or feed information to it, or what? (Tell me about it?)

5

J31. What kinds of things did <u>you</u> do while operating or helping to run this equipment?	J32. Did you use this (REPEAT NAME) almost constantly (AC), some of the time (ST), or very little (VL)?
_____	_____
_____	_____
_____	_____
_____	_____
_____	_____
_____	_____
_____	_____
_____	_____
_____	_____
_____	_____
_____	_____
_____	_____

──J33.

(IF MORE THAN ONE PIECE OF EQUIPMENT IS LISTED IN J29)

J34. CHECK ONE: | SOME OF THE EQUIPMENT | | NONE OF THE EQUIPMENT |
      (SEE Q.J32) | WAS USED ALMOST CONSTANTLY | | WAS USED ALMOST CONSTANTLY |

                      (SKIP TO Q. J40 PAGE 7)   (ASK Q.J35)

      J35. Thinking now of all the equipment that you operated or helped to run on
            your old job, did you use at least one piece or another of this equipment:

            | 1. Almost constantly | | 2. Some of the time | | 3. Very little |

6

J36.   Was this equipment the kind where you had to control absolutely everything it did
yourself, or did it do some things automatically?

| 1. SOME AUTOMATIC FEATURES | 3. NONE HAD AUTOMATIC FEATURES (SKIP TO Q. J40, PAGE 7) | ☐ DON'T UNDERSTAND Q. OR DON'T KNOW WHETHER HAD ANY AUTOMATIC FEATURES |

> J36a. For example, did it automatically feed a new piece after another was
> finished, or perform two or more operations one after another without
> your doing something, or automatically control the flow of work, or
> signal when a process was complete, or anything like that?
>
> | 2. SOME AUTOMATIC FEATURES | 4. NONE HAD AUTOMATIC FEATURES (SKIP TO Q. J40, PAGE 7) | 8. DON'T KNOW (SKIP TO Q. J40 PAGE 7) |

J37.   What kinds of things did it do automatically? _____

J38.   Was any of this equipment you worked with connected up to any kind of tape or
numeric control unit, or to a computer? (Tell me about it.)

| 1. SOME WAS CONNECTED TO CONTROL OR COMPUTER | 5. NONE WAS CONNECTED TO CONTROL OR COMPUTER (SKIP TO Q. J39) |

J38a.   Do you happen to know whether the computer (control unit) kept track of time,
or counted things, or recorded things like speed, size, shape, or temperature,
weight or anything like that? ...What did it do? ...Anything else it did?

J39.   Was any of the equipment you worked with connected up to another piece of equipment
by conveyors, or pipes, or electrical connections, or something like that - so that
they worked together as a unit? (Tell me about it.)

7

J40. We are also interested in any equipment which you didn't operate, or operated very little, but which was important for your work on your old job. I mean, for example, equipment you inspected, cleaned, designed, made, sold, installed, or repaired, or perhaps equipment in other departments or companies. Was there any such equipment that was important to your work on your old job?

| 1. YES | | 5. NO | → (SKIP TO Q.J44 BELOW)

J40a. What kinds of equipment or machinery were these? Any other kind?

(ASK J41-J43 ABOUT EACH PIECE OF EQUIPMENT)

J41. What was it called? (TRY TO GET EXACT NAME)	J42. What did it do?	J43. In what way was this equipment important to your work?

J44. Thinking again of <u>all</u> equipment important to your job, <u>including</u> any equipment that you have operated yourself, would you say the <u>greatest change</u> in equipment for <u>you</u> happened when you changed jobs, or since you have been on your present job? (I mean the change which had the greatest effect on you and your job.)

1. WHEN CHANGED JOBS	2. SINCE HAS BEEN ON PRESENT JOB	5. NO CHANGE IN EQUIPMENT EITHER TIME
	(GO TO "A" SCHEDULE, Q.D7, PAGE 14)	(SKIP TO PART II, Q. J59, PAGE 9)

J45. CHECK ONE: ☐ The big "A" is checked at the top of the Schedule A face sheet. (R's present job is affected by some equipment.)   ☐ The "A" is <u>not</u> checked (SKIP TO Q.J57, PAGE 9)

J46. Thinking of the greatest change in equipment you had - would you say that what <u>you</u> had to do on your job was <u>changed</u> by it, or did you keep on doing nearly the same thing in nearly the same way, or were the <u>jobs</u> so different you can't even compare them?

1. WORK ON JOB WAS CHANGED	3. JOBS SO DIFFERENT CAN'T BE COMPARED	5. KEPT ON DOING NEARLY THE SAME WORK
		(SKIP TO PART II, QJ59, PAGE 9)

J47. In what ways was your (work changed/job different)?

8

J48. Did you have to learn anything new or did you acquire any new skills when you started to work with the new equipment?

|1. YES| |5. NO|————→(SKIP TO Q. J56 BELOW)

J49a. How did you acquire the new skill or knowledge -- did you learn it by yourself on the job?

.T49b. Did someone train you on the job?

J49c. Or did you take a formal training program or course?

> CHECK AS MANY BOXES AS APPLICABLE, THEN ASK CONTINGENT QUES- TIONS SEPARATELY FOR EACH TRAINING PROGRAM

☐ FORMAL TRAINING PROGRAM OR COURSE -- --- ASK Q's J50-J55

☐ TRAINED ON JOB --- ASK Q's J54-J55.

☐ LEARNED BY SELF -- ASK Q. J55.

J50. Where did you take this training program or course -- did the company give it, or a union, did you go to a school or university, take a correspondence course, or what?

(IF COMPANY) J50a. Was this at the company you left or at the company you went to?

J51. Was this voluntary on your part or were you asked to do it?

J52. Who paid for it, or was it free?   PAID FOR OR PROVIDED BY: |1. RESPONDENT|

☐ FREE |2. OLD EMPLOYER| |3. NEW EMPLOYER|

J52a. Who provided it? |4. EQUIPMENT PRODUCER| |5. GOVERNMENT|

OTHER: _____

J53. Did you start this course or training program before you left your former company, or after you joined the new company, or in between?

|1. BEFORE| |5. AFTER| |3. IN BETWEEN; AT SAME TIME|

J54. How long a program or course was it -- how many hours or days or weeks?

_____ HOURS ALTOGETHER | IF ANSWER IN TERMS OF DAYS OR WEEKS, PROBE TO FIND TOTAL NUMBER OF HOURS

J55. What kinds of things did you learn? _____

_____

(IF DID NOT TAKE FORMAL TRAINING PROGRAM OR COURSE -- SEE Q. J49c ABOVE)

J56. At the time you changed over to the new equipment, did you have the opportunity to take a formal training program of some kind offered by a company, by a union or by the government?

_____

_____

9

J57. Altogether what helped you the most in getting used to the change in your work? (Anything else?)

_____

_____

_____   ____

J58. Was there anything which made it hard for you to get used to the change? (What was it?)

_____

_____

_____

PART II: CHANGE IN JOB CONTENT

Next we want to discuss with you some of the details of how your new job compares with the old job we have been talking about.

J59. Is there more physical work on the new job, or less, or is there no difference?

| 1. MORE | | 5. LESS | | 3. NO DIFFERENCE |

J60. Is the speed with which you have to work greater on the new job, or slower, or is there no difference?

| 1. GREATER | | 5. SLOWER | | 3. NO DIFFERENCE |

J61. Is more skill required of you on the new job, or less skill?

| 1. MORE SKILL | | 5. LESS SKILL | | 3. NO DIFFERENCE |

J62. Did the physical surroundings at work - things like light, heat, noise, space, ventilation - become more pleasant with the new job, or less pleasant?

| 1. MORE PLEASANT | | 5. LESS PLEASANT | | 3. NO DIFFERENCE |

J63. On the new job, do you have occasion to talk more often with other people at work, or are you more alone?

| 1. MORE OFTEN | | 5. MORE ALONE | | 3. NO DIFFERENCE |

(SKIP TO Q. J64, PAGE 10)

> J63a. Do you find this more pleasant, or less pleasant, or does it make no difference?
>
> | 1. MORE PLEASANT | | 5. LESS PLEASANT | | 3. NO DIFFERENCE |

10

J64.   Do you have more reports to write or more records to keep on the new job, or less, or is there no difference?

| 1. MORE | 5. LESS | 3. NO DIFFERENCE | 0. NOT APPLICABLE |

> J64a. Why is that? _____
>
> _____
>
> _____

J65.   Is there more danger of personal injury on the new job, or less danger?

| 1. MORE DANGER | 5. LESS DANGER | 3. NO CHANGE |

J 66. On the new job, is the opportunity to learn new things about the work greater than before, or less?

| 1.GREATER OPPORTUNITY | 5. LESS OPPORTUNITY | 3. NO CHANGE |

J67.   Is there more need for planning, judgment, or initiative on your part on the new job, or less need?

| 1. MORE NEED | 5. LESS NEED | 3. NO DIFFERENCE |

> J67a. In what ways are the jobs different in these respects (planning, judgment, initiative)? (Anything else?) _____
>
> _____
>
> _____

J68.   In some jobs, how your work is organized depends mostly on you, yourself. In others it's pretty much determined by the equipment you work with, by a production line, or by a number of people who have to work together. On your new job do you feel that you have more influence on how your work goes, or less influence than you did on your old job?

| 1. MORE INFLUENCE | 5. LESS INFLUENCE | 3. NO CHANGE |
| | | (SKIP TO Q. J69) |

> J68a. Why is it that you have (more/less) influence? _____
>
> _____
>
> _____

11

J69. If you made a mistake on your work, would it likely be more serious on the new job, or less serious?

| 1. MORE SERIOUS | 5. LESS SERIOUS | 3. NO CHANGE |

J69a. Why is that? _____

_____

(ASK ONLY IF R IS ENGAGED IN PRODUCTION WORK)

J70. Would you say that on your new job the quality of the product you make is more dependent on how well you do your work, or less dependent?

| 1. MORE DEPENDENT | 5. LESS DEPENDENT | 3. NO DIFFERENCE |

(ASK EVERYONE)

J71. Would you say that what you do on your job is more closely supervised on the new job, or less closely supervised compared to the old job?

| 1. MORE CLOSELY SUPERVISED | 5. LESS CLOSELY SUPERVISED | 3. NO CHANGE |

J72. What about the chances of unemployment or being laid off -- is the work more steady on the new job, or is it less steady?

| 1. MORE STEADY | 5. LESS STEADY | 3. NO CHANGE |

J73. For a person like yourself, are the chances of moving up to a better job greater with your new employer, or are they smaller?

| 1. GREATER CHANCE | 5. SMALLER CHANCE | 3. NO DIFFERENCE |

J73a. Why is that? _____

_____

12

J74. Would you say that in general your work is more interesting on the new job, or is it more monotonous?

1. MORE INTERESTING; LESS MONOTONOUS	5. LESS INTERESTING; MORE MONOTONOUS	3. NO CHANGE

J74a. Why is that? _____

_____

_____

##

(GO TO "A" SCHEDULE, Q. G2, PAGE 23)

##

Appendix II

SAMPLING METHODS AND
SAMPLING VARIABILITY

A. The Sample

The data for this monograph were obtained by selecting a representative national cross-section of households and interviewing all *labor force participants* in these households.[1] This is the same approach which is used by the U.S. Bureau of the Census in the Current Population Survey to obtain a sample of the labor force. In this study, the sample includes all household members who said they were working full-time or at least 20 hours a week at the time of the interview. Those who were unemployed at the time of interview and those who were working less than 20 hours were interviewed only if they stated that they would have liked to work at least 20 hours and would have been able to do so if work had been available. To be eligible for an interview, these people also must have worked 20 hours or more a week at some time during the past 5 years.

---

[1] In this report the term national refers to the 48 coterminous states (excluding Alaska and Hawaii) and the District of Columbia.

For the way in which labor force participation was ascertained see Appendix I above, especially questionnaire page 1.

A household includes all persons living in a dwelling. The Survey Research Center uses the dwelling unit concept defined by the United States Bureau of the Census, U.S. Census of Housing: 1950; Vol. I., "General Characteristics, Part I: U. S. Summary," page XVI. Dwelling units on military reservations are excluded from the study universe. Also excluded are persons living in nondwelling unit quarters; examples of these are: large rooming houses, residential clubs, dormitories, hospitals, and penal institutions.

For general discussion of sampling procedures see L. Kish and I. Hess, "The Survey Research Center's National Sample of Dwellings," Institute for Social Research, The University of Michigan, Ann Arbor, Michigan, 1965, ISR No. 2315.

The area probability sampling procedure used to select households progressed through several stages. First, a probability sample of 78 counties or county groups (primary sampling units) was chosen. The 78 primary sampling units included each of the 12 largest standard metropolitan areas which were self-representing in the sample. The remaining 66 primary sampling units represented 66 relatively homogeneous strata of about equal size. These strata were formed on the basis of the following criteria: geographic location; SMSA classification; size of the largest city, rate of population growth; major industry or major type of farming; in the South, proportion of nonwhite population. From each of the 66 strata, one county (or county group) was selected with probability proportional to population.

Instead of independent selections within each stratum, controlled probability selection was introduced for a more efficient sample. Within each of the four geographic regions the selections of primary areas were linked by a procedure which controls the distribution of sample areas by states and degree of urbanization beyond the controls effected through the stratification of primary sampling units. This controlled selection yields a more balanced sample and further reduces the sampling variance of characteristics.

Within each selected primary unit about five places, on the average, were chosen by probability methods. These places were cities, towns, suburban and rural areas. The selection of places within primary sampling units involved substratification by degree of urbanization and economic level. In the third stage of sampling, urban segments (blocks) or small segments of rural areas were chosen, again using stratification and random selection within strata. Finally, a sample of dwellings, in clusters of about four, was drawn from the segments.

Where more than one family occupied a chosen dwelling, each family in the dwelling was considered as falling into the sample. Within the family, all labor force participants (as defined above) were to be interviewed. In this manner a sample of about 3,400 eligible respondents was selected at an overall rate of 1/16,525. Repeated calls were made to obtain interviews with all eligible respondents. If after repeated calls a designated respondent was not at home, was ill, refused to be interviewed, or for some other reason the interview was not obtained, no substitution was made. Neither was the survey data adjusted for nonresponse. From the selected respondents 2,662 interviews were obtained, a response rate of about 78 percent.

B. Sampling Variability

Estimates from properly conducted sample interview surveys are subject to errors arising from several sources. Among these are sampling errors and response, reporting, and processing errors. A sample of 2,662 cases is not large, particularly when it comes to studying or comparing subgroups of the population. The moderate size of the sample puts a limit on analysis possibilities and gives rise to sampling errors; yet it has advantages also. Since this is a study which attempts to break new ground in the kinds of data collected, a crucial consideration in designing it was accuracy of reporting of new and difficult material. Careful interviewer training, close contact with the field staff, and careful coder training reduce reporting error. They are greatly facilitated when the sample is not too large.

Although both nonsampling and sampling errors are important in evaluating the accuracy of the data, measurement of each type is not always possible from the survey itself. But in the case of probability designs, measures of sampling variability can be calculated from the sample data. The discussion below is limited to sampling variability. In Section C we shall compare some important sample characteristics with independent estimates.

Percentages—Sample statistics reflect the random variations arising from interviewing only a fraction of the population. The distribution of individuals selected for a sample generally differs by an unknown amount from that of the population from which the sample is drawn. The value that would be obtained if interviews were taken by the same survey procedures with the entire population will be referred to as the population value. If different samples were used under the same survey conditions, some of the estimates would be larger than the population value and some would be smaller. The sampling error is a measure of the deviation of a sample statistic from the corresponding population value, but it does not measure the sampling variability of a particular sample estimate. The sampling error leads to the construction of an interval or range, on either side of the sample estimate, that includes the population value in a specified proportion of cases in a large number of samplings.

As used here, the term sampling error means two standard errors; it is the range, on either side of the sample estimate, chosen to obtain the 95 percent level of confidence. If a greater degree of confidence is required, a range wider than the two standard errors can be used. On the other hand, most of the time the actual error of sampling will be less than two standard errors; in about 68 percent of cases, a range of one standard error on either side of the sample estimate includes the population value.

For example, the survey estimate (Table 2-1) that 42 percent of the labor force operate equipment "almost constantly" is subject to a sampling error of about 2.8 percentage points. Thus the statement that the range 39.2 to 44.8 percent includes the population value would be true for at least 95 of every 100 samples drawn like the one for this study. The chances are that in 5 of every 100 samples the population value would be outside that range;

however, the chances are that in 68 of each 100 samples, the population value would be included in the range 40.6 to 43.4 percent, the estimate plus and minus one standard error.

The sampling error of the proportion of the labor force having a certain characteristic depends on the size of the sample and also on the size of the proportion being estimated. Approximately, the sampling error is inversely proportional to the square root of the sample size. Thus, the sampling error of an estimate based on 400 respondents is about one-half as large as that of an estimate based on 100 respondents.

Standard errors also vary with the proportion being estimated and reach a maximum, for samples of a given size, when the proportion is 50 percent. (However, the relative size of the error decreases as the size of the percentage increases.) The relation of sampling error to sample size and the proportion being estimated is evident in the formula for the computation of sampling errors for simple random samples. The sampling error of such a sample is equal to $2\sqrt{[p(1-p)]/(n-1)}$, where **p** is the proportion under consideration and **n** is the sample size. Although the survey uses a complex rather than a simple random sample, the relationship of sampling error to sample size and the proportion being estimated is somewhat similar to that of the preceding formula.

There are other important factors that influence the size of the sampling error of any characteristic based on interviews from the entire sample or from some specific subgroup. Stratification at several stages of sampling tends to reduce sampling error while clustering of the sample in a limited number of counties, cities, blocks and rural areas tends to increase sampling variation. The joint effect of such factors varies for every type of estimate and for every subgroup of the population. The fact that sampling errors in this study are frequently higher than simple random sampling errors arises because dwellings and respondents were sampled in clusters, a procedure that may increase sampling error if the characteristic being sampled also occurs in clusters. The measures of sampling errors presented in Tables II-1 and II-2 take into consideration the particular characteristics of the complex sample design used for this study. The sampling errors were *not* derived on assumptions of simple random sampling.

The sampling errors themselves are products of the sampling processes and are subject to the effects of random fluctuations as well as to the effect of sample design. Estimates of sampling errors are presented in Table II-1.[2] The figures are *average* values. Some survey statistics may have higher sampling errors while others may have lower sampling errors. Statistics subject to

---

[2]For computational formulas and procedures used to summarize estimates of sampling errors, see Kish and Hess, *op. cit.* pp. 43-53.

TABLE II-1

APPROXIMATE SAMPLING ERRORS OF PERCENTAGES[a]

Estimated	Number of interviews									
percentages	2500	1500	1000	700	500	400	300	200	100	50
50	2.9	3.5	4.0	4.6	5.4	5.9	6.8	8.2	11	16
30 or 70	2.8	3.2	3.7	4.3	5.0	5.5	6.2	7.5	10	15
20 or 80	2.4	2.8	3.2	3.7	4.3	4.8	5.4	6.6	9.1	13
10 or 90	1.8	2.1	2.4	2.8	3.2	3.6	4.1	4.9	6.9	10
5 or 95	1.3	1.5	1.8	2.0	2.4	2.6	3.0	3.6	5.0	7.0

[a]The figures in this table represent two standard errors. Hence, for most items the chances are 95 in 100 that the value being estimated lies within a range equal to the estimated percentage plus and minus the sampling error.

higher than average sampling variability are those for some subgroups of the population—for example, geographical regions, metropolitan and rural areas, movers, and Negroes. However, for many practical decisions the approximations presented in the table will be satisfactory. If more precision is required, the sampling error could be calculated for the specific statistic under investigation.

Differences—Differences between survey estimates are often of even greater interest than the levels of the estimates. These differences reflect the random fluctuations of the sampling process as well as differences in population values. The sampling errors of differences may be used to determine the range that would include the true difference between population values of two compared classes in a given proportion of trials. As with sampling errors of single percentages, greater or lesser degrees of confidence in a statement are associated with larger or smaller multiples of the standard error.

Estimates of *average* sampling errors of differences are presented in Table II-2. To illustrate the use of the table, consider the proportions of men and women who operate equipment "almost constantly" (Table 2-1). The number of cases in the first group is around 1,800; the number of cases in the second group is around 850. The table shows 40 percent of men as opposed to 47 percent of women operating equipment "almost constantly," a difference of 7 percentage points. If we look in Table II-2 for percentages between 35 and 65 and for numbers of cases of 1,800 and 850, we see the difference required for significance at the 5 percent level is about 5.3 percentage points. Thus the difference of 7 percentage points between the two groups is significant (at two standard errors).

TABLE II-2

APPROXIMATE SAMPLING ERRORS OF DIFFERENCES[a]

| For percentages from 35% to 65% | | | | | | | | | |
Interviews	2000	1500	1000	700	500	300	200	100	50
2000	4.4	4.7	5.1	5.6	6.3	7.5	8.8	12	16
1500		4.9	5.3	5.8	6.4	7.6	8.9	12	16
1000			5.7	6.2	6.8	7.9	9.2	12	17
700				6.6	7.1	8.2	9.5	12	17
500					7.7	8.7	9.9	13	17
300						9.6	11	13	17
200							12	14	18
100								16	19
50									23

| For percentages around 20% and 80% | | | | | | | | | |
	2000	1500	1000	700	500	300	200	100	50
2000	3.5	3.7	4.1	4.5	5.0	6.0	7.0	9.5	13
1500		3.9	4.3	4.6	5.1	6.1	7.2	10	13
1000			4.6	4.9	5.4	6.3	7.3	10	13
700				5.3	5.7	6.6	7.6	10	13
500					6.1	7.0	7.9	10	14
300						7.7	8.5	11	14
200							9.3	11	14
100								13	16
50									18

| For percentages around 10% and 90% | | | | | | | | | |
	2000	1500	1000	700	500	300	200	100	50
2000	2.7	2.8	3.1	3.4	3.8	4.5	5.3	7.1	9.8
1500		3.0	3.2	3.5	3.9	4.6	5.4	7.2	9.9
1000			3.4	3.7	4.1	4.8	5.5	7.3	10
700				3.9	4.3	4.9	5.7	7.4	10
500					4.6	5.2	5.9	7.6	10
300						5.8	6.4	8.0	10
200							7.0	8.5	11
100								10	11
50									14

| For percentages around 5% and 95% | | | | | | | | | |
	2000	1500	1000	700	500	300	200	100	50
2000	1.9	2.0	2.2	2.4	2.7	3.3	3.8	5.2	7.1
1500		2.1	2.3	2.5	2.8	3.3	3.9	5.2	7.2
1000			2.5	2.7	2.9	3.5	4.0	5.3	7.2
700				2.9	3.1	3.6	4.1	5.4	7.3
500					3.3	3.8	4.3	5.5	7.4
300						4.2	4.7	5.8	7.6
200							5.1	6.1	7.9

[a]The values shown are the differences required for significance (two standard errors) in comparisons of percentages derived from two different subgroups of the population.

C. Comparisons With Independent Estimates

Besides calculating sampling errors, the representative nature of the sample may be assessed by comparing some basic sample characteristics with corresponding estimates from the Current Population Survey of the U. S. Bureau of the Census. If we are willing to allow for some differences in the definition of labor force membership (the Census figures include people who work only 1-19 hours, while the present survey covers only people who work at least 20 hours) and in the wording of questions, then we may accept as meaningful the comparisons of age, education, and sex distributions shown in Tables II-3, II-4, and II-5.

The age distributions in Table II-3 agree quite well. Although the sample for this study appears somewhat older than the Census sample, differences could be attributed mainly to sampling variability. The sampling error of the Census estimates would be much smaller than those shown in this report (Tables II-1 and II-2), because the Census estimates are derived from about 50,000 households and consequently have a higher level of precision than those obtained from the present survey. Furthermore, the Survey Research Center sample excluded students who worked less than 20 hours a week.

The education distributions in Table II-4 agree less closely. The major discrepancy is the higher percentage of people with a high school degree and the correspondingly lower percentage with 1 to 3 years of college in the Census distribution. This may be largely a matter of definitions. The Survey

TABLE II-3

PERCENTAGE DISTRIBUTION OF LABOR FORCE BY AGE, 1967

(Survey Research Center and Census Comparison)

Age groups	Survey Research Center estimate	Census estimate
Under age 25	13.1%	16.5%
25-34	21.5	20.2
35-44	24.3	22.9
45-54	23.1	21.9
55-64	14.8	14.5
Age 65 or older	3.2	4.0
Total	100.0%	100.0%

Source: Census estimates as published in Table D, p. A-9, Special Labor Force Report No. 92, "Educational Attainment of Workers, March 1967", United States Department of Labor, Bureau of Labor Statistics.

TABLE II-4-

PERCENTAGE DISTRIBUTION OF LABOR FORCE BY YEARS OF SCHOOL COMPLETED, 1967

(Survey Research Center and Census Comparison)

Years of school completed	Survey Research Center estimate	Census estimate
0-7 grades	8.1%	10.1%
8-11 grades	29.1	29.5
High school degree	32.6	36.6
Some college; 1 to 3 years	15.5	11.8
College, 4 years	9.0	7.5
College, 5 years or more	5.7	4.5
Total	100.0%	100.0%

Source: Census estimates as published in Table D, p. A-9, Special Labor Force Report No. 92, "Educational Attainment of Workers, March 1967", United States Department of Labor, Bureau of Labor Statistics.

TABLE II-5

PERCENTAGE DISTRIBUTION OF LABOR FORCE BY SEX, 1967

(Survey Research Center and Census Comparison)

Sex	Survey Research Center estimate	Census estimate
Male	67.7%	63.7%
Female	32.3	36.3
Total	100.0%	100.0%

Source: Census estimates as published in Table D, p. A-9, Special Labor Force Report No. 92, "Educational Attainment of Workers, March 1967", United States Department of Labor, Bureau of Labor Statistics.

Research Center classifies anyone with some college work as having "some college;" the Census Bureau leaves people who have not completed one *full* year of college in the category "high school degree." In addition the Survey Research Center estimate of college graduates exceeds the Census estimate by a small, but statistically significant, margin; the opposite is true for the estimate of labor force members with less than 8 years of schooling.

A comparison of the sex distributions in Table II-5 reveals that there is a somewhat higher proportion of men in the Survey Research Center sample than in the Census sample. This difference is to be expected in view of the fact that the present survey excluded people working less than 20 hours (unless they were unemployed). Many of the labor force participants with very short hours are women.

Appendix III

DEFINITION AND MEASUREMENT
OF AUTOMATION LEVEL*

The equipment which members of the labor force operate could be classified according to a number of characteristics. Of particular relevance for this study was a classification according to degree of automation. In Section A below the conceptual framework of the automation classification is described. Since many respondents were unable to provide enough information to code the eleven-fold classification outlined, the scale actually coded is a condensed version of the one originally visualized. The full conceptual scheme should nevertheless be of interest to others who may be working with different kinds of data. The implementation of the conceptual scheme is discussed in Section B.

A. The Conceptual Scheme[1]

There are many definitions of the word "automation,"[2] yet there would be some agreement that the concept of control is central to the study of

---

*The conceptual scheme outlined here was worked out by John A. Sonquist and Richard L. Crandall at The University of Michagin. Roger Hybels and Bruce Van Dyke, also at The University of Michigan, were responsible for adapting the conceptual scheme to the data collected in the Survey and were in charge of the actual coding process.

[1]The conceptual scheme is derived from the work of James R. Bright, *Automation and Management,* Norwood, Massachusetts, The Plimpton Press, 1958.

[2]For a discussion of varied definitions, see M. Lefton, "On Defining Automation," *Outlook,* Vol. 2, No. 4, Western Reserve Univ., Cleveland, 1965.

automation. Control involves decision making: starting, stopping, guiding the amount, direction, and duration of all operations performed by the equipment.

The most elementary form of control is, of course, *manual control* of equipment which is not power assisted. Hand controlled power tools remove some trivial control functions from the human operator. Yet, this type of equipment does not involve significant mechanization of control over the amount, direction, and duration of the operations performed.

A second form of control may be termed *fixed mechanical.* A single operation or a fixed sequence of operations is set up in advance. The operation(s) may be initiated by the operator or by inputs of materials. The introduction of materials-handling equipment, such as a conveyor usually is regarded as one important feature of automation. If several pieces of equipment operate synchronously and are actuated by the introduction of a piece of material transferred between them on a conveyor without human intervention in the process, then operator functions are further reduced. Control remains at the fixed mechanical level, but the span of the implementation of mechanical control is extended. The equipment is, therefore, at a somewhat higher level of automation than individual pieces of mechanically controlled equipment, but does not involve logical control. Equipment may be very complex and yet be controlled by a combination of manual and fixed mechanical methods.

Equipment characterized by manual or fixed mechanical control as the primary mode of operation can be broken down into single or multi-system machines, depending on whether a single or a series of operations are performed by the machine. Fixed mechanical equipment can be subdivided further into single or multi-cycle machines. The former perform one or a series of operations and then return to an idle position. The latter perform one or a series of operations repetitively. In either case the process is initiated by the operator or by an input of material, but is carried forward by the machine.

Control, in the sense of making decisions about what to do next, is not complete without information about the characteristics of the operations being performed. In both the manual and fixed mechanical modes of control, the operator performs important functions with respect to information and how it is used. In the manual mode, continuous monitoring of discrete operations is required, sometimes aided by gauges, dials, etc. In the fixed mechanical mode operation sequences are possible without the necessity of operator intervention. However the differences in the operator's information processing and decision-making functions are those of degree rather than kind. Like equipment under complete manual control, fixed mechanical equipment cannot *change* its own sequence of operations on the basis of information gathered during the processing. Once a sequence of operations is started, it continues unless the operator intervenes.[3]

The third type of control may be termed *logical.* Equipment of this type is characterized by its capability of being "programmed." This means specifically that the equipment has a repertoire of operations it can perform, that these operations can be executed by the equipment itself in an arbitrary order upon recognition of the necessary signal, and that information for producing the requisite signals can be recorded and stored in a form "readable" by the equipment itself.

Equipment characterized by logical control can be further broken down according to the amount of feedback and also the flexibility incorporated into the device. Feedback involves the use of sensory units to provide information about the material (or information) that is being processed. These data are used by the equipment itself, without human intervention, to modify or change the operations being performed. Flexibility is the capability of performing many types of operations in different sequences. Its two components are the number of discrete operations that can be performed and the extent to which they can be modified and sequenced in different ways to do different kinds of things. In general, a "programmable" machine is flexible since its operation repertoire may be arranged in almost any order.

The more complex machines may combine several methods of control. In that case the most meaningful classification appears to be by the most automated method of control (i.e. highest on the scale), which must however be of some importance. Control functions which are peripheral to the operation of the equipment or to the work of the operator should not be used in classifying the equipment. For example, even highly automated machines may have to be started manually by the operator. Or again, the automatic shut-off triggered by overloads or jams involves a "feedback" device. This form of logical control is almost universally present on equipment where such a disturbance is frequent and would cause damage. Yet it should not determine the overall classification of the equipment.

The scheme is depicted in Chart I. It may be noted that this classification is applicable not only to equipment which performs operations on "things" (solids, liquids, gases) but also to equipment which processes information. Moreover, reading from top to bottom, the classification provides an ordering of the level of automation (although the ordering of categories 5 and 6 as well as 9 and 10 may be open to argument). The eleven detailed classifications can be combined to form a smaller number of more comprehensive, but still ranked, classifications.

---

[3]An automatic shut-off for overload conditions is often present on equipment *primarily* controlled by fixed mechanical means. Nevertheless the primary control mode does not involve the ordinary use of this type of feedback circuit.

CHART I

AUTOMATION LEVEL CLASSIFICATION
OF EQUIPMENT

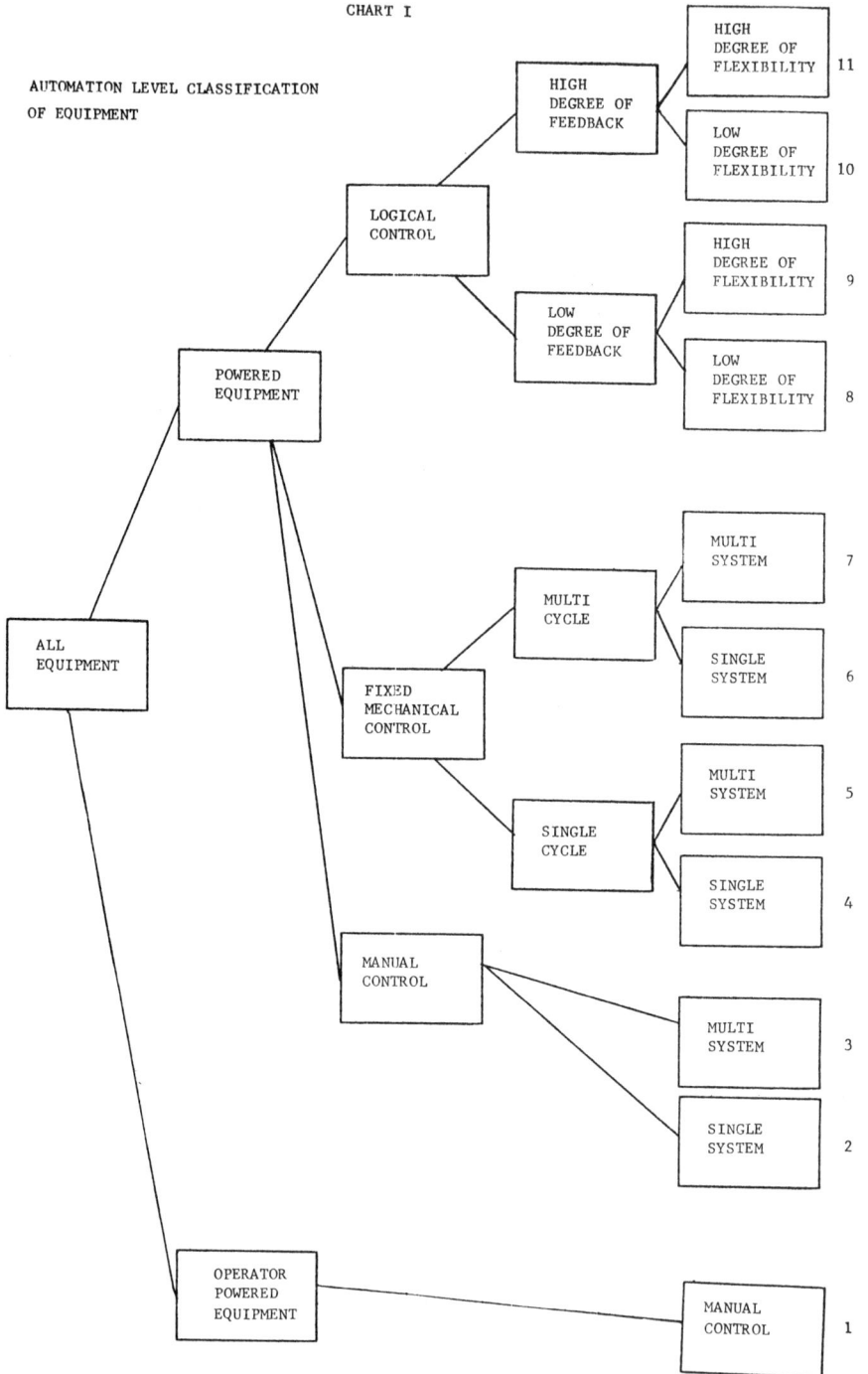

Using a scheme like this, it is possible to classify all of the equipment used by a member of the labor force. Alternatively, it is possible to identify only the highest level of equipment associated with the particular person's job.

B. *Implementation*

The questionnaire was designed with a view toward constructing the automation scale just described. It was recognized at the outset, however, that some of the subcategories would have to be collapsed, because it might not be possible to obtain all the required information from respondents. Also, it was recognized that some of the subcategories might not contain enough cases to make them useful in the analysis. With these considerations in mind, it appeared advisable, after the completed interviews were studied, to build a six-point scale.

Manual control was divided into three classes: Operator powered and controlled; powered, single-system; and powered with multiple systems. For the most part, operator powered and controlled machines are hand tools but include others, i.e. paint brush, broom and manual typewriter. The powered, single-system manually-controlled machines include electric drills, precision inspection tools, vacuum cleaners, power lawn-mowers and air hammers among others. The category of powered multi-system manually controlled machines was an attempt at a classification which would include complex machines still requiring a man for their main control. The classic example for this is the automobile and other vehicles used in commerce and production, such as bull-dozers and tractors. It also includes fork lifts, and cranes, as well as conventional switchboards. As can be seen by the examples, man remains the prime guider and controller of the operation of this equipment.

The fixed mechanical category could not be readily subdivided. It includes a large variety of machines, among them much of the mass production and assembly line equipment. Once started, these machines repeat the same operation or sequence of operations and produce identical products until the process is completed or a malfunction develops. The manner of operation and the specifications of the product are set mechanically into the machine by the use of cams, templates, dies, and the setting of gears, tensions, lengths, etc. Once these factors are "fixed," the machine can do one operation or a sequence of operations. A few examples are: duplicating equipment, cash registers, milling, grinding, and packaging machines, punch presses, automatic drillers, gas pumps, textile looms, and two-way radios. It is evident that man remains the starter, overseer, and often stopper of the operation.

For machines with logical control subsclassification again was difficult. However, computers were the most important kind of equipment in this category. It was decided to lump tape controlled machines and computers together. The number of cases of tape control was very small, and this category for all practical purposes is composed of computers. Any logically controlled machines not falling into the computer category were put into "other logical control."

A few machines were difficult to fit into this set of criteria. In large part, these were research instruments and biomedical machinery. Such equipment as heart-lung machines, electrocardiographs, electronic test instruments and P-H meters fell into this category. Most of them were put into the fixed mechanical category.

For each respondent, equipment which he operates directly and equipment operated by others but important for his work are coded separately. Many respondents worked on an assembly line or conveyor but used power tools and the like as their major equipment. In these cases the power tool was coded as the equipment they operate and the assembly line (fixed mechanical control) as equipment which they do not operate, but which is important for their work. In those cases where the respondent operated several pieces of equipment, the machine ranking highest on the automation scale was the one that was coded. The same was done for equipment where the contact was indirect. The distribution of cases along the automation scale is shown in Table 2-3 Part B. As is indicated there, the "operator powered and controlled" and the "other logical control" categories contain very few cases. For purposes of analysis these categories were therefore combined with the adjacent ones. Table 2-3 Part B also shows that less than 2 percent of cases were uncodable either for lack of information or because the scale did not fit.

The automation scale was coded by graduate students trained in engineering. In an effort to determine the uniformity of the coding, a cross—check was carried out. Three coders independently coded the respondent's own equipment on his present job for a total of 716 interviews. Disregarding discrepancies involving adjacent categories, more significant errors (two or more categories apart) were found in 77 cases, or 10.7 percent of the time. However, since two-check coders could find the same error of the third coder, one mistake might well result in two errors being listed. To that extent the 10.7 percent would be an overstatement of errors. On the other hand, the 716 interviews included over 40 percent with respondents who operated no machines at all; in those cases no error was likely to arise.

Appendix IV

THE MULTIPLE CLASSIFICATION ANALYSIS

The principal statistical method used in Chapter 5 and throughout this monograph is the Multiple Classification Analysis (MCA), a simple extension of multiple regression analysis.[1] Regression analysis is essentially a method for predicting variations between individuals in a dependent variable making use of information about these same individuals on a number of other factors (independent or predictor variables) simultaneously. The estimating procedure has the characteristic that the summed squared deviations of the actual from the predicted values of the dependent variable are smaller than for any other equation using the same independent variables and the same additive model.

The new feature of MCA is that it converts each subclass of the independent variable into a set of "dummy variables," which take the value one if an individual belongs to a particular subclass of each factor, and zero if he does not. Instead of a single regression coefficient for a numerical variable like age, we have a set, one attached to each age group. Dummy variable regression techniques are particularly needed for independent variables which are unordered (occupation or industry). Even when the independent variable is a numerical measure such as income or age, little precision is lost when it is converted to a number of classes. Thus MCA requires of the data only that they can be grouped into exhaustive and mutually exclusive classes—no assumptions concerning scaling need be made.

---

[1] For a further description of this technique, see James N. Morgan, et al., *Income and Welfare in the United States,* New York, McGraw-Hill, 1962, pp. 508-511; also Frank Andrews, James N. Morgan and John Sonquist, Multiple Classification Analysis, Survey Research Center, University of Michigan, Ann Arbor, Michigan, 1967.

MCA also requires no assumptions about the kind of function which relates the dependent to the independent variables. That is, rather than assume linearity, or some restricted quadratic or log form, MCA is entirely flexible. It allows the data to determine the shape of the relationship between some independent or predictor variable and the dependent variable. For example, MCA could reveal that the chance of becoming unemployed is similarly affected by being young and old, but quite differently affected by being middle-aged. MCA can reveal the nature of non-linearities without any a priori assumptions.

The program output shows how each independent variable relates to the dependent variable, both before and after adjusting for the effects of other independent variables. MCA is unique in that the weighted sum of subclass regression coefficients for each independent variable equals zero. Thus the actual means of the dependent variable for each subclass of each independent classification can be presented in the form of deviations from the grand mean.[2] These are termed "unadjusted deviations." When these deviations are corrected for the effects of other independent variables, they are termed "adjusted deviations."

The unadjusted deviations can tell us whether people who work with highly automated equipment earn incomes above or below the average for the labor force as a whole and by approximately how much. The adjusted deviations indicate whether this difference persists in an analysis that also takes account of education, occupation, and other factors. In addition, the differences between the unadjusted and the adjusted deviations provide evidence about the extent to which intercorrelations among the independent variables affected the unadjusted deviations.

One limitation of MCA is the assumption of additivity which, however, is inherent in any multiple regression procedure. That is, it is assumed that each predictor variable affects the dependent variable in an independent manner, regardless of the values of other predictor variables. Where interactions are known to exist, the data can be reduced to an additive model. Different subgroups of the sample can be analyzed separately if it is suspected that the explanatory variables act differently in these subgroups. In this study, workers with equipment change and those with job change were sometimes analyzed separately. A second method of handling interactions may be used where two factors interact but do not affect other explanatory variables. In such a situation the two interaction variables can be combined. In the analysis in Chapter 5 and 6, job change, transfer, and equipment change and their combinations were treated as a single explanatory variable.

---

[2] This attribute distinguishes MCA from dummy variable regressions in which the class coefficients are normally deviations from the mean of an arbitrarily excluded class. See D.B. Suits, "Use of Dummy Variables in Regression Equations," *Journal of the American Statistical Association,* 1957.

A disadvantage of MCA vis-a-vis conventional regression analysis is that with MCA, statistics of significance for the explanatory variables cannot be readily derived. MCA routinely computes a multiple correlation coefficient which indicates the magnitude of relationship between the dependent variable and all predictors considered together. One procedure for assessing the significance of individual predictors is to reanalyze the data, adding each independent variable in turn to an otherwise standardized set of independent variables, in order to see the amount by which the total explanatory power of the regression is increased. This expensive procedure is used to a limited extent in this monograph.

In addition, a beta coefficient is produced for each predictor variable, which is analogous to the partial beta coefficient in ordinary multiple regression using numerical variables.[3] The beta coefficient is invariant with respect to the standard deviation of the dependent variable and is adjusted for the variability of the predictor. It hence provides an index of the relative importance of each independent variable in their joint explanation of the dependent variable. The one difference from ordinary beta coefficients is that the beta coefficients from the MCA program will all be positive, because the different directions of effect were already accounted for by the signs of the adjusted deviations.

---

[3]For the precise deviation of the beta coefficient from the MCA program see Andrews, Morgan and Sonquist, *op. cit.,* pp. 117-123.

BIBLIOGRAPHY

Listed below are books and articles published or prepared in 1968 by the staff of the Economic Behavior Program of the Survey Research Center.

Barfield, Richard and James N. Morgan. *Early Retirement: The Decision and The Experience,* 1969.

Dunkelberg, William C. and Frank P. Stafford. The cost of financing automobile purchases. *Review of Economics and Statistics,* 1969.

Katona, George, James N. Morgan, and Richard E. Barfield. Retirement in prospect and retrospect. *Trends in Early Retirement* (Occasional Papers in Gerontology No. 4). Ann Arbor: The University of Michigan Institute of Gerontology, March 1969, 27-49.

Katona, George and Eva Mueller. *Consumer Response to Income Increases* (An Investigation Conducted in the Year of the Tax Cut). Washington, D.C.: Brookings Institution, 1968.

Katona, George. On the Function of Behavioral Theory and Behavior Research in Economics. *American Economic Review,* LVIII, March 1968, 146-150.

Katona, George. Consumer Behavior: Theory and Findings on Expectations and Aspirations. Proceedings, *American Economic Review,* LVIII, 2, May 1968, 19-30.

Katona, George. Consumer Behavior and Monetary Policy. In *Geldtheorie und Geldpolitik* (Festschrift for Guenter Schmoelders). Berlin, Germany: Duncker and Humbolt, 1968, 117-132.

Lansing, John B., Charles Wade Clifton, and James N. Morgan. *New Homes and Poor People.* Ann Arbor: Institute for Social Research, 1969.

Morgan, James N. Family Use of Credit. *Journal of Home Economics, 60,* January 1968.

Morgan, James N. Some pilot studies of communication and consensus in the family. *Public Opinion Quarterly, 32,* 1 Spring 1968, 113-121.

Morgan, James N. The supply of effort, the measurement of well-being, and the dynamics of improvement. *American Economic Review, 58,* May 1968.

Morgan, James N. Survey analysis: applications in economics. In *International Encyclopedia of the Social Sciences, 15,* New York: Macmillan, 1968, 429-436.

Sirageldin, Ismail Abdel-Hamid. *Non-Market Components of National Income,* 1969.

Sonquist, John A. Problems of getting sociological data in and out of a computer. Paper read at the American Sociological Association, Boston, August 1968, 22 p.

Stafford, Frank P. Concentration and labor earnings: comment. *American Economic Review, 58,* 1, March 1968, 174-181.

Stafford Frank P. Student family size in relation to current and expected income. *Journal of Political Economy,* 1969.

Data collected by the Economic Behavior Program are available on either punched cards or computer tapes, together with a detailed code describing the content of the cards or tapes. Thus, interested scholars or other parties may obtain or prepare further analysis beyond that presented in this volume.

SURVEY RESEARCH CENTER PUBLICATIONS

Survey Research Center publications should be ordered by author and title from the Publications Division, Department L, Institute for Social Research, The University of Michigan, P.O. Box 1248, Ann Arbor, Michigan 48106.

1960 Survey of Consumer Finances. 1961. $4 (paperbound), 310 pp.

1961 Survey of Consumer Finances. G. Katona, C. A. Lininger, J. N. Morgan, and E. Mueller. 1962. $4 (paperbound), $5 (cloth), 150 pp.

1962 Survey of Consumer Finances. G. Katona, C. A. Lininger, and R. F. Kosobud. 1963. $4 (paperbound), 310 pp.

1963 Survey of Consumer Finances. G. Katona, C. A. Lininger, and E. Mueller. 1964. $4 (paperbound), 262 pp.

1964 Survey of Consumer Finances. G. Katona, C. A. Lininger, and E. Mueller. 1965. $4 (paperbound), 245 pp.

1965 Survey of Consumer Finances. G. Katona, E. Mueller, J. Schmiedeskamp, and J. A. Sonquist. 1966. $4 (paperbound), $6 (cloth).

1966 Survey of Consumer Finances. G. Katona, E. Mueller, J. Schmiedeskamp, and J. A. Sonquist. 1967. $4 (paperbound), 303 pp.

1967 Survey of Consumer Finances. G. Katona, J. N. Morgan, J. Schmiedeskamp, and J. A. Sonquist. 1968. $5 (paperbound), $7 (cloth), 343 pp.

1968 Survey of Consumer Finances. G. Katona, W. C. Dunkelberg, J. Schmiedeskamp, and F. P. Stafford. 1969. $5 (paperbound), $7 (cloth), 287 pp.

Automobile Ownership and Residential Density. John B. Lansing and Gary Hendricks. 1967. $3, 230 pp.

The Geographical Mobility of Labor. John B. Lansing and Eva L. Mueller. 1967. $6.50, 421 pp.

Multiple Classification Analysis. James N. Morgan, John A. Sonquist and Frank M. Andrews. 1967. $3

Productive Americans: A Study of How Individuals Contribute to Economic Progress. James N. Morgan, Ismail Sirageldin, and Nancy Baerwaldt. 1966. $5, 546 pp.

Residential Location and Urban Mobility: The Second Wave of Interviews. John B. Lansing. 1966. $2.50 (paperbound), 115 pp.

Private Pensions and Individual Saving. George Katona. 1965. $1.50 (paperbound), $2.50 (cloth), 114 pp.

Consumer Behavior of Individual Families Over Two and Three Years. Richard F. Kosobud and James N. Morgan (Editors). 1964. $5 (paperbound), $6 (cloth), 208 pp.

Residential Location and Urban Mobility. John B. Lansing and Eva Mueller. 1964. $2 (paperbound), 142 pp.

Residential Location and Urban Mobility: A Multivariate Analysis. John B. Lansing and Nancy Barth. 1964. $2 (paperbound), 98 pp.

The Travel Market, 1964-1965. John B. Lansing. 1965. $4 (cloth), 112 pp.

*The Changing Travel Market. John B. Lansing and Dwight M. Blood. 1964. $10 (cloth), 374 pp.

The Detection of Interaction Effects. John A. Sonquist and James N. Morgan. 1964. $3 (paperbound), 292 pp.

The Geographic Mobility of Labor, a First Report. John B. Lansing, Eva Mueller, William Ladd, and Nancy Barth. 1963. $3.95 (paperbound), 328 pp.

*The Travel Market 1958, 1959-1960, 1961-1962. John B. Lansing, Eva Mueller, and others. Reprinted 1963 (originally issued as three separate reports). $10, 388 pp.

*The Travel Market 1955, 1956, 1957. John B. Lansing and Ernest Lillienstein. Reprinted 1963 (originally issued as three separate reports). $10, 524 pp.

*Package of three available for $25.00.

Location Decisions and Industrial Mobility in Michigan, 1961. Eva Mueller, Arnold Wilken, and Margaret Wood. 1962. $2.50 (paperbound), $3 (cloth), 115 pp.

OTHER BOOKS BY MEMBERS OF THE ECONOMIC BEHAVIOR PROGRAM

Transporation and Economic Policy. John B. Lansing. Free Press, 1966.

The Mass Consumption Society. George Katona. McGraw-Hill, 1964.

Income and Welfare in the United States. J. N. Morgan, M. H. David, W. J. Cohen, and H. E. Brazer. McGraw-Hill, 1962.

An Investigation of Response Error. J. B. Lansing, G. P. Ginsburg, and K. Braaten. Bureau of Economic and Business Research, University of Illinois, 1961.

The Powerful Consumer. George Katona. McGraw-Hill, 1960.

Business Looks at Banks: A Study of Business Behavior. G. Katona, S. Steinkamp, and A. Lauterback, University of Michigan Press, 1957.

Consumer Economics, James N. Morgan, Prentice-Hall, 1955.

Contributions of Survey Methods to Economics. G. Katona, L. R. Klein, J. B. Lansing, and J. N. Morgan. Columbia University Press, 1957.

Psychological Analysis of Economic Behavior. George Katona. McGraw-Hill, 1951. (Paperback edition published in 1963.)

Economic Behavior of the Affluent. Robin Barlow, H. E. Brazer, and J. N. Morgan. Washington, D.C.: Brookings Institution, 1966.

Living Patterns and Attitudes in the Detroit Region. John B. Lansing and Gary Hendricks. A report for TALUS (Detroit Regional Transportation and Land Use Study), 1967, 241 pp. (Available only from TALUS, 1248 Washington Blvd., Detroit, Mich. 48226–$5 to nongovernmental agencies.)